Models for Biomedical Research

A NEW PERSPECTIVE

Committee on Models for Biomedical Research
Board on Basic Biology
Commission on Life Sciences
National Research Council

NATIONAL ACADEMY PRESS
Washington, D.C. 1985

NATIONAL ACADEMY PRESS 2101 CONSTITUTION AVENUE, NW WASHINGTON, DC 20418

NOTICE: The project that is the subject of this report was approved by the Governing Board of the National Research Council, whose members are drawn from the councils of the National Academy of Sciences, the National Academy of Engineering, and the Institute of Medicine. The members of the committee responsible for the report were chosen for their special competences and with regard for appropriate balance.

This report has been reviewed by a group other than the authors according to procedures approved by a Report Review Committee consisting of members of the National Academy of Sciences, the National Academy of Engineering, and the Institute of Medicine.

The National Research Council was established by the National Academy of Sciences in 1916 to associate the broad community of science and technology with the Academy's purposes of furthering knowledge and of advising the federal government. The Research Council operates in accordance with general policies determined by the Academy under the authority of its congressional charter of 1863, which establishes the Academy as a private, nonprofit, self-governing membership corporation. The Research Council has become the principal operating agency of both the National Academy of Sciences and the National Academy of Engineering in the conduct of their services to the government, the public, and the scientific and engineering communities. It is administered jointly by both Academies and the Institute of Medicine. The National Academy of Engineering and the Institute of Medicine were established in 1964 and 1970, and respectively, under the charter of the National Academy of Sciences.

The study reported in this publication was conducted at the request of, and funded by, the National Institutes of Health under Contract No. NO1-RR-3-2112.

Library of Congress Catalog Card Number 85-60945

International Standard Book Number 0-309-03538-4

Front and rear cover illustrations from *De Motu Animalium* by Alphonsi Borelli. Ex Typographia Anglei Bernabo. Romae MDCLXXX. Courtesy, Dibner Room, Special Collections Branch, Smithsonian Libraries. Washington, D.C.

Printed in the United States of America

NAS NAE IOM NRC National Academy Press

The National Academy Press was created by the National Academy of Sciences to publish the reports issued by the Academy and by the National Academy of Engineering, the Institute of Medicine, and the National Research Council, all operating under the charter granted to the National Academy of Sciences by the Congress of the United States.

Dedication

This report is dedicated to George Streisinger (1927-1984). He chaired our agenda subcommittee, organized the development workshop, and in many other ways left his imprint on the work of this committee. As a scientist, as a colleague, and as a citizen, George radiated a sense of wisdom, honor, and care for his fellow man. We profoundly missed his good-humored and penetrating insights during our final deliberations. The entire scientific community will miss George Streisinger for many years to come.

ERRATA

p. vii, para. 4, lines 2-3, should read "... The discussion of the broader concept forced us ..."

p. viii, para. 1, line 2, replace "rosophila" with "*Drosophila*"

p. 99, fourth workshop participant and affiliation from top should read:

Alan Gelperin AT&T Bell Laboratories Murray Hill, N.J.	Cellular and System Level Studies of Associative Learning in Protohumanoid Molluscs

Models for Biomedical Research: A New Perspective

ISBN 0-309-03538-4

Committee on Models for Biomedical Research

HAROLD J. MOROWITZ, *Chairman*
Department of Molecular Biophysics and Biochemistry
Yale University
New Haven, Connecticut

J. WOODLAND HASTINGS, *Vice Chairman*
Biological Laboratories
Harvard University
Cambridge, Massachusetts

MICHAEL V. L. BENNETT
Department of Neuroscience
Albert Einstein College of Medicine
Bronx, New York

KENNETH B. BISCHOFF
Department of Chemical Engineering
University of Delaware
Newark, Delaware

VINCENT J. CRISTOFALO
Wistar Institute of Anatomy and Biology
Philadelphia, Pennsylvania

W. GARY FLAMM
Toxicological Sciences
Bureau of Foods
Food and Drug Administration
Washington, D.C.

LAWRENCE I. GILBERT
Department of Biology
University of North Carolina
Chapel Hill, North Carolina

JOSHUA LEDERBERG
Office of the President
Rockefeller University
New York, New York

KEITH R. PORTER
Department of Biological Sciences
University of Maryland
Catonsville, Maryland

GORDON SATO
W. Alton Jones Cell Science Center
Lake Placid, New York

GEORGE STREISINGER*
Institute of Molecular Biology
University of Oregon
Eugene, Oregon

SAMUEL WARD
Department of Embryology
Carnegie Institution of Washington
Baltimore, Maryland

National Institutes of Health Project Officer

JAMES D. WILLETT
Division of Research Resources

National Research Council Staff

WALTER G. ROSEN, Study Director
FRANCES M. PETER, Editor
GLADYS E. COUCH, Secretary

*Deceased, August 1984

Preface

This report constitutes the response by the Committee on Models for Biomedical Research to a charge from the National Institutes of Health (NIH) to assemble, review, evaluate, and report on the relevance of various models for biomedical research sponsored by the NIH. The evaluation includes invertebrates and nonmammalian vertebrates, *in vitro* techniques (cell and tissue culture systems), and nonbiological approaches (e.g., mathematical and computer-assisted systems).

The committee realized it could not survey the 2 million or so known species to identify systems of highest priority for the study of the major unanswered questions about normal and pathological processes in humans. Since each species is a potential model, the task clearly required some unifying ideas or methodological taxonomy.

The committee approached its task by holding a series of workshops to explore the various ways some of the country's most able scientists selected organisms or nonbiological models to study specific questions or, conversely, how they selected questions to learn more about the model of interest. The participants in these workshops were chosen for their standing in their respective research areas, their breadth of knowledge in their disciplines, and their ability to provide a clear presentation. The workshops thus became a tutorial in which 53 presenters educated the committee in a wide area of model usage at the forefront of modern biology. I would like to express our gratitude to the workshop participants for the clarity of their presentations and the vigor with which they joined in discussions of our charge. Our sample of the research community made up in quality what it may have lacked in size.

As the workshops unfolded, their contents allowed us to narrow our gaze and to focus more systematically on the concept of "model" in various experimental disciplines. The discussion of the broarder concept forced was to verbalize what had been a tacit principle — the overwhelming importance of the "unity within diversity" of biology that is often stated as a theme, but is seldom extensively developed, in introductory textbooks. What was preeminently clear from the content of the workshops is that the unity goes far beyond the basic biochemical level and is seen at every hierarchical tier. The interspecific transfer of information, a key step in surrogate modelling, is inherent in the extensive series of relationships that tie together individual species or other biological entities in a complex, highly connected body of information and understanding.

In looking at the areas covered, it is clear that there are ways of performing necessary biological functions that, once evolved, appear to remain through evolutionary history. Since we have few laws and lack a firm understanding of some of the processes involved in interspecies correlations, the relationships are often empirical. This collection of relationships among processes and structures across the taxa is the raw material of organismic modelling in its most general sense. All sorts of structures and functions of a given organism are analogous to similar structures and functions in many other organisms, which thereby constitute potential models for the subject of interest.

To make the preceding more tangible, consider recent studies on mapping a mouse DNA sequence that is homologous to a gene sequence involved in *Drosophila* morphogenesis. Here, species separated by hundreds of millions of years of evolutionary history appear to utilize common develomental mechanisms. Thus in a narrow sense rosophila can be a model for the mouse, or in our extended view, the developmental information across all of biology provides a complex of highly interrelated patterns upon which various aspects of mouse development can be analyzed and modelled.

One is led to speculate that there are as yet undiscovered laws of morphogenesis and physiological signalling that will explain the many high order overlaps of molecular structures discussed in the workshops. Even in the absence of such laws, however, the surprising overlaps provide opportunities for a wealth of empirical generalizations that support our concept of modelling as a complex interconnection rather than the one-to-one concept of classical models. The latter is a special case of the former and can now be properly seen in that context. With these ideas in place, we are able to respond to the charge of examining the relevance and limitations of various classes of models.

A project of the magnitude of the present one can only come into being in a relatively short time with the help and cooperation of a substantial number of individuals. The committee would first like to acknowledge the help of Project Officer James Willett, National Institutes of Health, and our consultants Kenneth Schaffner, University of Pittsburgh; Evelyn Fox, Northeastern University; and John Jacques, University of Michigan, who provided us with insights in areas of their special expertise. The workshops came together through the efforts of organizers and keynote speakers Charles DeLisi, Jesse Roth, Bela Julesz, Paul Gross, William Wood, and John Farrar.

The continuing work of the entire enterprise was carried out by the National Research Council staff under the guidance of Walter G. Rosen. Dr. Rosen made this undertaking a labor of love. Without his constant reminding, cajoling, and gentle needling, I feel that this report might well have been published in 2001. His contributions were, however, more than just administrative, for very often he was able to supply a germinative idea or a key reference when we were floundering. Indeed his contribution to our thinking was fully that of a committee member. As the project entered the writing stage, the firm but fair editorial hand of Frances Peter converted masses of handwritten, typed, and printer-generated materials into a coherent document. We wish to acknowledge our deep thanks to Mrs. Peter and Dr. Rosen, and to Ms. Gladys Couch for her patient clerical support.

HAROLD MOROWITZ
Chairman
Committee on Models for
Biomedical Research

Contents

Models for Biomedical Research

A NEW PERSPECTIVE

1

Executive Summary

Biomedical research encompasses a vast array of activities that have as their ultimate objective improved understanding of the human organism in health and disease. Like other experimental sciences, biomedical research owes much of its success to the investigation of systems that serve as models, i.e., analogs, of different aspects of human structures and functions. Historically, living organisms of all kinds, and preparations derived from them, have served as experimental systems in biomedical research, although for many biomedical problems, particularly those most directly related to diseases, models have been sought primarily from among other vertebrates. This committee in its charge was asked to focus on nonvertebrate systems.

This study was conducted in response to a request from the National Institutes of Health (NIH) to the National Academy of Sciences to assemble, review, evaluate, and report upon the relevance and limitations of biomedical research models. The committee that performed this study within the Commission on Life Sciences of the National Research Council evaluated *in vitro* (cell and tissue culture) systems, nonmammalian vertebrates, invertebrates, protists, and nonbiological models.

Because the various subdisciplines of biomedicine are so numerous, and the number of species is so large, a systematic, comprehensive evaluation of all potential models for all areas of biomedical research was precluded. Instead, the committee decided to sample the spectrum of biomedical research modelling by adopting a case-study approach. Five topics, each covering an important area of basic biomedical research, were selected for examination in a series of case-study workshops. Through these workshops, the committee examined and evaluated experimental models for the investigation of cellular immunology, regulation, learning, diseases and aging, and development. A sixth workshop was subsequently arranged to examine mathematical modelling in biomedical research.

Workshop participants were asked to present critical reviews of their immediate research fields, giving particular attention to the use of models. In discussions following the individual presentations, the committee further explored general issues regarding the advantages and limitations of various models.

During the course of its study, the committee noted that the term model is used with several different meanings in biomedical research and that these usages are frequently not defined or specified. Models in both the physical and biomedical sciences are analogs. That is, the models possess the same or similar structures or functions as the system of interest, and information derived from the model is presumed to apply in some measure to the system of interest. Models are useful in research to the extent that they are good analogs. They are used because they may be simpler, more accessible, and in biomedicine may involve more acceptable organisms from a number of points of view, not all strictly scientific.

Because of biological evolution, biomedical research has an additional analytical tool at its disposal -- homology, which is correspondence in structure and function derived from a common evolutionary origin, i.e., a common gene sequence. The most closely related species, or its systems, is generally presumed to offer the best homologs. Phylogeny, however, is not always known in detail, and unresolved problems concerning evolutionary events and pathways are numerous. Functionally similar biological systems may be due to evolutionary (genetic) relatedness or, alternatively, to convergent evolution, where the structures or systems evolved independently. Care must therefore be used in evaluating the degree of homology and the extent to which it relates to analogy.

Some biological mechanisms, such as the coding of genetic information and the pathways of metabolism, arose early in evolution. These mechanisms have been highly conserved and are widely shared by organisms, including humans, at the cellular and molecular levels. Thus, good analogs for these fundamental molecular mechanisms in humans can be found in a wide array of organisms, some of which, such as bacteria, have structures and functions far less complex than those of mammals.

The workshops demonstrated that the results of biomedical research can be viewed as contributions to a complex body, or matrix, of interrelated biological knowledge built from studies of many kinds of organisms, biological preparations, and biological processes at various levels. On the basis of this research, a series of generalizations has evolved within and between these levels of organization in the matrix and between taxa. Useful information is transferred across all phyla. With the development of this network view come insights into further connections among data, enabling investigators to, among other things, make many valid extrapolations to normal and disease processes in humans.

This concept of a matrix of biological knowledge arose repeatedly in the committee's discussions. Ideally the matrix is an encyclopedia of biological knowledge organized by our most general theories of unity within diversity. It is extensively cross-referenced by principles and generalizations, rather than alphabetically.

The direct transfer of information from an animal model or biological preparation to the human can be viewed as one-to-one modelling, i.e., modelling from one group of organisms, such as humans, to another that has several analogous factors of interest. The committee concluded, however, that biomedical research is usually more inclusive, i.e., many-to-many modelling to the matrix of biological knowledge, in which many features at various hierarchical levels can be modelled at each level to the range of taxa in which the features appear. (See Chapter 3 for a more detailed discussion of one-to-one and one-to-many modelling.)

One-to-one modelling is subsumed within many-to-many modelling and is especially useful in modelling disease states and higher integrated organismal activities. Thus, the most effective modelling strategy in the study of complex functions at higher levels, i.e., systems or organisms, often begins with analysis of component functions and the identification of models for components of those functions, reflecting the common reductionist approach used in experimental research. The transfer of information from simple models to components of a complex process is sometimes difficult, because higher levels of organization in organisms may display properties not seen in their component parts.

The workshops illustrated, and perhaps dictated, many-to-many modelling to the matrix of biological knowledge as a conceptual framework for this study. They also demonstrated that individual investigators are often more interested in acquiring a general understanding of a process, or function of a structure, than in the direct transfer to humans of the knowledge gained from the study. The primary objective is to understand the process per se and then to establish connections between the process in one species to the same or related processes in other organisms.

Thus, the workshops provided many examples of organisms serving as sources of information about the fundamental nature of processes that are found throughout the living world and made understandable through addition of new information to the matrix of biological knowledge. Following are a few illustrations from the many discussed at the workshops.

- Regulation of cell-cell interactions has played a role in normal development, the immune response, the inflammatory response, and learning. Models for various cell-cell interactions are found in protists, sponges, marine worms, cell and tissue culture systems of many taxa, and mathematical formulations.

- Although some aspects of learning can be observed in microorganisms and invertebrates and described at the molecular level, others at present can be studied only in intact primates or in tissues removed from them.

- Important progress in understanding certain human metabolic diseases has resulted from studies on models of components of the diseases that can sometimes be found in a broad range of nonmammalian organisms or in cell and tissue culture systems.

- Cell and tissue culture studies to examine the cellular basis of immunology have provided insights unattainable through research on intact organisms.

Nonmammalian organisms and cell and tissue cultures have been and will be used as models for biomedical research and clinical medicine. They cannot entirely replace intact mammals, however, since the validity of the interspecific transfer of information must be experimentally verified.

Although the value of mathematical formulations in describing biological processes is unquestioned, the committee was less certain about the extent to which practical results would be obtained from purely theoretical approaches. There is a communications gap between mathematical modellers and experimentalists that is readily acknowledged by both groups. Important advances could undoubtedly come from closing this gap.

In the concept of the matrix of biological knowledge -- a multidimensional matrix of data and constructs -- there is little or no demarcation between basic biology and its applications in human medicine. Data from biomedical research form a web of interconnections among the structures and functions of all organisms, from the simplest microbes to the human and other primates. Although other primates, the organisms phylogenetically closest to humans, may be required for research on certain aspects of human biology, important insights have been gained from observations on much different or simpler organisms.

The universal genetic code was first discovered and studied in bacteria and viruses but is now known to be shared by all living things, including human beings. This is only one of a very large number of cases demonstrating the genetic and functional interrelatedness of all living things resulting from their common evolutionary origin. To fulfill its mission most effectively, therefore, the NIH should support research on a variety of organisms and systems, including invertebrates, microorganisms, and plants, selected for their potential for decisive experimentation that will lead to development of biological principles. The potential for applying the results to human medicine is great. When investigators seek a simple model for a specific medical problem, their success may depend on having available a large variety of known and understood organisms or systems from which to choose.

Early in its deliberations, the committee decided that it could not perform its task effectively without reference to vertebrates, because the evaluation of nonvertebrate models and model systems requires a basis for comparison. Both poikilothermic vertebrates (i.e., fishes, amphibians, and reptiles) and homeothermic vertebrates (i.e., birds and mammals) constitute important sources of models for some areas of biomedical research. For some problems, such as those involving the integration of complex physiological systems, organisms from these groups are essential, and no equally satisfactory models are to be found among nonvertebrate organisms or preparations.

The most appropriate model for some investigations will often be a vertebrate. Indeed, the development of new constructs and theories from work on invertebrates, microorganisms, cell and tissue cultures, and theory (mathematical modelling) may generate, at least transiently, a need for using more vertebrates in defining the problem and in validation experiments.

Computer storage and analysis of biological data and relationships should facilitate identification of potential models and model systems as well as important connections between different compartments in the matrix. The research required to process the data in the matrix for computer storage and retrieval should in itself reveal such relationships.

The problem-centered format of the workshops stimulated exchanges among investigators using different models to study the same or closely related phenomena. The focus on models, their applications, and limitations gave rise to new perceptions of the interrelatedness among the various fields of research. Intersystem comparisons among investigators conducting varied NIH-supported investigations might be expected to increase effective use of model systems in a broad spectrum of biomedical research.

Workshop participants generally believed, and the committee agreed, that the NIH should support promising research proposals in biomedicine and leave the selection of the model to the insight of the investigator. Many pointed out, however, that present procedures for allocating resources may limit support to work on too few species.

There is an unpredictable element in the discovery of new and better models and model systems. Thus, general support of studies on the biology of invertebrates and other organisms can be expected both to add to the fund of basic biological knowledge in the matrix and to provide, from within the network, improved biomedical research models.

The committee concluded that the NIH can best contribute to progress across the spectrum of biomedical research by the following actions:

- encouraging investigators and managers to think of models not necessarily as analogs relating directly to humans on a one-to-one basis but, rather, as potential sources of information generalizable to the total body of biological knowledge;

- encouraging the development of new systems with novel features by supporting research on nonmammalian species, including representatives of taxa that have not previously been well studied;

- encouraging continuing strong support for research on a small number of intensively studied organisms that are accessible to comprehensive genetic, molecular, behavioral, and developmental analysis;

- encouraging the development of detailed knowledge about various aspects of the biology of nonmammalian organisms to discover their connections to the rest of the matrix;

- encouraging the application to and rigorous testing of mathematical modelling in experimental biomedical research;

- supporting good research without taxonomic or phylogenetic bias, including support for comparative and phylogenetic studies; and

- continuing to support research using the best mammalian models where adequate nonmammalian models are unavailable.

2

Approach to the Study

During the course of its study, the Committee on Models for Biomedical Research evaluated intact nonmammalian organisms, *in vitro* techniques, such as cell tissue and organ culture, and nonbiological approaches, such as mathematical and computer modelling. The committee interpreted its charge from the National Institutes of Health (NIH) broadly to include the following questions:

- How can information be transferred from taxon to taxon most usefully with the ultimate goal of transference to *Homo sapiens* in normal and pathological states?

- What is the degree of confidence in this interspecific information transfer? This question relates to the extent to which a general biology (in this context a structure analogous to theoretical physics that subsumes particular cases within general laws) exists, and the extent to which biologists have succeeded in formulating it.

- How are the problems inherent in information transfer related to the levels of organization (e.g., molecules, cells, tissues, organs, or organisms) of the phenomena under study?

The committee regarded these questions as basic to its task of devising more specific definitions and to its study of the applications of models.

The views of the committee changed considerably as a result of the workshops and the ensuing discussions, which led to an evolving perception of its charge. The directions that were chosen and the strategy of analysis were modified by what was learned. In a very real sense, the committee was forced to inquire into the nature of biological understanding, theory formation in biology, and the conceptual structure of biological science. The nature of the problem led the group to confront some issues usually considered by philosophers of science rather than by experimentalists. This was dictated by the analysis and the nature of the issues.

At the outset, the committee realized that the array of subjects of biomedical interest was far too vast to examine comprehensively, considering the constraints of time and resources. It decided instead to examine the issue by a case-study workshop approach, sampling five research areas that promised to provide useful methodological insights. A sixth workshop, on mathematical modelling, was added because this methodology was insufficiently treated in the first five. In order to gain the most current information in each of these areas, experts were asked to present their views and to provide an information base in their specialty in workshops organized by biological function rather than by taxa. Subsequently the workshops were analyzed across a range of functional issues by taxonomic level and methodology used. This approach, which examines biological functions across the taxa, provided the most comprehensive view that could be taken in a limited study of this nature.

During the course of the study, workshops on the following six topics were held: cellular immunology, biological regulation, learning, diseases and aging, development, and mathematical modelling.

CELLULAR IMMUNOLOGY

Knowledge about the immune response is of great value in understanding disease. The classical immune system used in this area of research is restricted to chordates -- a fairly narrow taxonomic domain. The very complexity of the intact organism often precludes direct study of individual components of the immune system. However, through cell and tissue cultures, one can examine the responses of clones, mixed clones, and other combinations of cells in *in vitro* experiments, free of confounding effects found at higher levels of organization. The primary issue in this area is not intertaxonomic transfer but, rather, extrapolation from *in vitro* experiments to improve the understanding of *in vivo* processes.

BIOLOGICAL REGULATION

This workshop was focused on an area of physiological research that began in the last century with Claude Bernard's concept of a milieu interieur (Bernard, 1865). Here again a range of lower taxa and regulatory mechanisms were examined to determine the extent to which a general physiology or general biochemistry of homeostasis exists. Regulation is a very central theme in biology and biomedical research and was therefore an obvious choice for workshop discussion.

LEARNING

Workshop participants discussed a variety of organisms from procaryotes to humans and a range of techniques from biochemistry to computer analysis and applied mathematics. The program was designed to address the following questions: Do common structures and processes for learning exist across a broad taxonomic range? What are the common biological features of a learned response?

DISEASES AND AGING

This subject was chosen because of its close relationship to the mission of the NIH. The committee asked scientists studying certain pathological and aging processes to explain how information from model systems is most useful in furthering understanding and in developing treatment strategies for gerontological disorders and for certain selected disease states.

DEVELOPMENT

This workshop was designed to use phylogenetic breadth and experimental accessibility to examine the early stages of morphogenesis, to explore the underlying cellular and molecular mechanisms, and to identify common features. Particular emphasis was placed on induction and repression of gene expression.

THEORY (MATHEMATICAL MODELLING)

Unlike the case-study workshops, which were organized around particular biological problems or processes, participants examined the application of mathematical formulations to a variety of biological processes, ranging from the molecular level to the organismal level. Theory as used here refers to mathematical and computer models, which must be verified by comparison with biological systems.

THE WORKSHOP FORMATS

The case-study workshop participants were selected because their research lies at the frontier of their field and exemplifies new or extended use of models or model systems. Each workshop opened with an introductory talk by a leading researcher, who presented an overview of the field to be explored. This was followed by presentations of 8 to 10 papers, each 30 to 40 minutes in length. The participants were asked to present critical reviews of their specialty areas, giving special attention to the models or model systems used, their strengths and limitations, and their potentials for further development. Brief discussions followed each presentation. In a separate discussion session at the end of each workshop, participants and committee members were encouraged to comment and speculate about the general use and limitations of models, to give examples of the use of various models to elucidate important medical problems, and to comment on the transferability of information among biological systems.

The invited speakers were asked to provide the committee with abstracts of their presentations, including key references. These abstracts were circulated to all members of the committee and were used by the workshop organizers to prepare the summaries of the workshops, which appear in Appendices B through F. Citations in the summaries are derived largely from those abstracts and are not intended to be comprehensive.

The workshops were announced to the public through a notice that appeared in Science on April 13, 1984. Announcements were also published in the newsletter of The Johns Hopkins Center for Alternatives to Laboratory Animal Testing and in other journals and newsletters. In addition, announcements were circulated internally at the NIH and were posted at the Marine Biological Laboratory in Woods Hole, Massachusetts, near the Academy's Summer Study Center where two of the workshops were held.

THE THEORETICAL STRUCTURE OF GENERAL BIOLOGY

Workshop participants were virtually unanimous in expressing a sense of intellectual stimulation as a result of being challenged to think rigorously about intertaxonomic and interhierarchical information transfer. Committee members found that the workshops broadened their awareness of the uses, potential uses, and limitations of various models and experimental systems, and caused some of them to revise their view of the implications of biological relationships between and among diverse taxa, other model systems, and human beings.

The view of biological science that emerged from the workshops and the committee deliberations may appear to some to be novel, but in fact is very close to the consensus view of experimental biologists since the firm establishment of evolutionary theory. It is, however, a consensus view that is not always made explicit. The position, expressed in the phrase "unity in diversity," has been a credo of biologists for more than a century. The diversity refers to the millions of species, past and present, each with morphological and physiological characteristics sufficiently unique to enable systematists to distinguish it from other species. The unity first emerged in common anatomical features and was strengthened by the universality of the cell theory and generalizations pertaining to the two great classes of cells: eucaryotes and procaryotes. Studies on morphogenesis revealed similar patterns covering a vast range of species. For example, the program of blastula formation followed by gastrula formation and the development of ectoderm, mesoderm, and endoderm is characteristic of all vertebrates and most invertebrates as well, and can be studied at the biochemical level across taxa (Gross, 1968).

The next step toward the concept of unity was the rise of biochemistry and the realization that there is a universal chart of intermediary metabolism (Sallach, 1972) involving a relatively small number (approximately 1,000) of intermediates. The intermediary metabolism of any species is a subset of this chart, which therefore has as much generality for biochemistry as the periodic table does for chemistry.

As molecular biology became a highly developed science, a whole new set of generalizations came into being: the universality of the genetic code, the ubiquity of ribosomes, and the uniformity of macromolecular synthesis apparatus. These features bind the living world within a tight conceptual framework at the cellular and biochemical level.

The committee's sense of the unity within biology was broadened during the workshops. The participants impressed upon the committee that at every hierarchical level from molecules to ecosystems, common hardware, common programs, and common strategies are used to achieve diverse ends. This general view is shown in the work of Jesse Roth and coworkers, demonstrating the ubiquity of peptide hormones. This was described at the workshop on regulation (see Appendix D).

Each of the workshops produced unexpected examples of intertaxonomic relationships. At the end of its study the committee viewed biology as a highly interrelated matrix of data, connections, and correlations. This matrix concept is discussed in detail in Chapter 5. The underlying evolutionary theme provides for homology, yet analogies without demonstrated homologies are also found in the matrix.

All these relationships and the data base that made them known constitute the theoretical structure of general biology, which is very different from the theoretical structure of physics. This difference and its implications for modelling are discussed in Chapters 3 and 5.

REFERENCES

Bernard, C. 1865. Introduction à l'Étude de la Médecine Expérimentale. J.-B. Baillière & Fils, Paris. 400 pp. (An Introduction to the Study of Experimental Medicine. 1927. H. C. Greene, transl. Macmillan, New York. 226 pp.)

Gross, P. 1968. Biochemistry of differentiation. Ann. Rev. Biochem. 37:631-660.

Sallach, H. J. 1972. Intermediary Metabolism Charts. P. L. Biochemicals, Inc., Milwaukee, Wisconsin.

3

What Is a Model?

> The problem of science will consist precisely in this, to seek the unitary character of physiological and pathological phenomena in the midst of the infinite variety of their particular manifestations.
>
> --Claude Bernard (1865, p. 124 in Eng. Trans.)

The concept of a model seems to have preceded the frequent appearance of the term in biomedical research literature. In his classic work, *An Introduction to the Study of Experimental Medicine*, Bernard (1865) discussed "the usefulness to medicine of experiments on various species of animals." (See English edition, pp. 122-129.) Krogh (1929) stated, "For a large number of problems there will be some animal of choice, or a few such animals, on which it can be most conveniently studied." Almost half a century later, this became known as the August Krogh principle (Krebs, 1975). Krebs and Krebs (1980) cautioned that "an uncritical application of this principle may lead to fallacious generalizations, because extrapolating findings from one species to another is not invariably valid." Ross (1981) carried this point further, arguing that comparative physiology, rather than achieving its objectives of contributing to knowledge of phylogenetic relationships and of discovering the origins of physiological functions, has in reality dealt with the description of adaptations. On the other hand, Bullock (1984) argued that "comparative neuroscience is likely to reach insights so novel as to constitute revolutions in understanding the structure, functions, ontogeny, and evolution of nervous systems." Without using the term, these authors were discussing what we refer to today as models.

The various kinds of models and their meanings were discussed by Ransom (1981), who wrote:

> In its simplest form, a model is a simplified representation of a structure.... A *heuristic* model is a model used to discover how a process works rather than being a *descriptive* model of the process.... The definition of a

> heuristic model is in fact rather simple but it is the way in which such models are constructed that gives rise to most difficulty in classification. Static models can be heuristic; for example, the form of a protein molecule is often worked out by trial and error construction of plastic models....

Ransom described a variety of heuristic modelling techniques, such as the following:

Paper and pencil (static) models. Ransom cited D'Arcy Thompson's (1917) _Growth and Form_ as a classic example of the application of this type of modelling to development. Thompson's grid technique analyzed growth in terms of localized asymmetries arising from differential growth rates during early development.

Mathematical models. In this type of modelling, mathematical equations are used to describe a process. The discrete (statistical or probabilistic) model is mentioned as another type of mathematical model. [A report prepared by a committee of the National Research Council (1981a) contains a discussion of mathematical models, including statistical models, simulation models, and both qualitative and semi-quantitative models.]

Computer models. Elements from both pencil and paper models and mathematical models can be combined to produce hybrid models, normally animated as _simulations_. Ransom proposes that a simulation is "the dynamic representation of a model on a computer." According to this author, "The sequential representation of a process at different states in time is the essential basis of the computer model...."

Substitute system models. Ransom includes the use of living organisms as models, pointing out the frequent advantage of providing simpler systems than the ones in which our interest might be centered. Used in this sense, the model is often referred to as a surrogate. Russell and Birch (1959) discuss fidelity and discrimination as factors governing the way in which the model differs from the original.

This report is not intended to be an exhaustive survey of either the uses of the term model or of the organisms, preparations, and mathematical procedures that have served as models or model systems. Studies that review the use or potential use of biological materials as model systems are fairly numerous, including several conducted by the National Research Council:

Mammalian Models for Research on Aging (National Research Council, 1981a);

Marine Invertebrates, a volume in the Laboratory Animal Management Series (National Research Council, 1981b);

The Future of Animals, Cells, Models, and Systems in Research, Development, Education, and Testing (National Research Council, 1977);

Animal Models of Thrombosis and Hemorrhagic Diseases (Department of Health, Education, and Welfare, 1976);

Psychopathology: Experimental Models (Maser and Seligman, 1977);

Animals and Alternatives in Toxicity Testing (Balls _et al_., 1983);

Species-Specific Potential of Invertebrates for Toxicological Research (Kaiser, 1980);

Trends in Bioassay Methodology: In Vivo, In Vitro and Mathematical Approaches (National Institutes of Health, 1981); and

Invertebrate Models in Aging Research (Mitchell and Johnson, 1984).

The proceedings of a symposium sponsored by the Society for Experimental Biology to examine the use of models and analogs in biology provides an illustration of the various applications of the term model in the life sciences (Beament, 1960). Included in the topics covered in this volume are mechanical models, electrical analogs, computers, kinetic models, models in cybernetics, psychological models, and educational models.

As used in biology, the concept of a model has not always been consistent and involves two broad classes, determined by whether the modelling is based on analogy or on homology. Modelling based on analogy is used extensively in the physical as well as in the biological sciences. Homology-based modelling appears to be uniquely biological, reflecting evolution and the genetic fixing of historical events into DNA sequences.

Modelling by analogy involves a point-by-point process relating one structure to another or one process to another (in mathematical terminology, mapping). This means finding correspondences with respect to some features. It requires that there be similarities between the two things being connected by the modelling relationship. Thus, for example, it is possible to model the concentration field in a diffusion problem by the electrostatic voltage field, provided that the geometries and boundary conditions are approximately set. This is possible because both phenomena follow the same differential equation.

All analog computers operate because the computer hardware elements exist in some kind of modelling relationship to the elements in the problem being solved. This accounts for the rigid, high specificity of analog computers and is the reason for their replacement by digital computers, which can model any mathematical structure, according to

Turing's theorem about the existence of a universal computer (Turing, 1936).

In naval architecture, studies using model basins are made possible by the same hydrodynamic equations, once the scaling factors have been taken into account. Here the structural features in common to the model and the object being modelled are quite apparent.

Both physics and engineering commonly use analog models. Indeed modelling, if we include similar mathematical features as one of its bases, is a major part of the activity of those sciences. We can therefore formalize the idea of relationships by analogy within the structure of physical sciences.

The idea of reasoning by analogy goes far back in the history of science. Kant (1790) wrote, "Analogy, in a qualitative sense is the identity of relations subsisting between grounds and consequences -- causes and effects -- so far as such identity subsists despite the specific differences of the things, as of those properties, considered in themselves (i.e., apart from this relation), which are the source of similar consequences." This idea has persisted in slightly modified form.

Analogies and models as they relate to the physical sciences have been reviewed by Achinstein (1968). He writes:

> In all of the cases considered we might describe the model or analogy as (or as containing) (1) a representation of X; but (2) one that is either not literal, or not faithful in all respects, or not complete, and may represent X in some "indirect" manner; and (3) one that utilizes something more or less familiar, known, understood, readily grasped, or easily experimented upon. Thus, a representational model represents X, but not completely and not necessarily literally, by utilizing something Y that is familiar or more readily grasped. In a theoretical model we represent X, but only approximately and not completely, by bringing it under, or at least utilizing parts of, some more basic theory or theories that are familiar and understood. In an imaginary model we represent X but not in a way intended to be literal, by imagining how X could satisfy certain conditions, where either the set of conditions or the way we represent X is more or less familiar and understood. In an analogy X is represented in an indirect way by being shown to be similar in some though not all respects to a distinct item more familiar or more readily grasped.

These notions have been presented in a somewhat different way by Margenau (1977). He presents what is in many ways a consensus view of how physics is methodologically structured at the most general level. Because of its generality, it applies to all of science, including biology. According to this view, science starts with observation, phenomena, sense perceptions, the raw material of our knowledge of the world. By rules of correspondence we now move to the existence of objects (reification) and the behavior of those objects. We then construct an elaborate set of theoretical devices: laws, definitions, theories, postulated entities (e.g., atoms and electrons), which are connected by logical and mathematical relationships. The validity of this structure is tested by the agreement of statements or predictions with the observed phenomenological world.

Margenau introduced the general term "constructs" to apply to all the theoretical conceptual devices listed above. Physical reality for him, which we generalize to scientific reality, consists of a cycle in which we continuously go from the world of observation through understanding by constructs back to observation.

Next consider two independent sets of observations on different entities. These can be connected insofar as some of the same constructs are used in the understanding of each of them. Analogies then exist between the two systems with respect to the overlapping constructs and the two systems model each other. The more frequently constructs are used in gaining an understanding of independent sets of observables, the stronger is the analogical relation between them.

Modelling by analogy is also used in the biological sciences. For example, since flight in bees, birds, and bats has certain aerodynamic features in common, a modelling relationship is possible (although the differences in these particular systems may be of more interest than the similarities). In the same sense modelling between a goose and an airplane is possible. Here the modelling can have unusual features. For example, both bird and airplane carry their fuel as saturated hydrocarbons, although it is esterified to glycerol in birds. This maximizes the energy-to-weight ratio of fuel in an oxidizing atmosphere.

Biology is characterized by a second type of modelling, i.e., modelling by homology, which seems unique to that field of study. The objects of biology -- organisms -- have an evolutionary history that is embedded in their genomes. As a result of evolution from a common origin, there are many shared genetic sequences and common functions between organisms. In general, species that have diverged most recently have the closest resemblances in DNA sequences and functions of protein and RNA derived from these sequences. The relationships between organisms resulting from their shared evolutionary history and matching DNA sequences form the basis of models by homology. Thus, for example, livers of the rat, pig, and human are homologously related in an evolutionary sense, and we would suspect and indeed do find similarities that make them suitable for an analogous modelling relationship. On the other hand, the bat wing, the foreleg of the horse, and the porpoise flipper are homologously related to the human arm, as revealed by comparative anatomy and the fossil record, but they provide poor analog models for human arm function.

Models by homology are thus of heuristic value in the search for analogs, but they become functionally useful only when they are also good models by analogy for the phenomenon or structure being studied. Thus, if one is investigating lipid solubilization in mammalian metabolism, the rat-pig-human liver homology transfers to a very useful analogous modelling system for research. If one is interested in flight, the bat-horse-porpoise homology may be of limited value for special questions.

The reason that homolog models sometimes fail to be good analog models is twofold. First, individual physiological adaptations may make homologs poor analogs. Ross (1981) illustrated this by describing the phylogenetic irregularities in the distribution of respiratory pigments in invertebrates.

Second, convergent evolution may make good analogs out of genetically very distant structures and processes. A commonly cited example for convergent evolution is the mammalian eye and the cephalopod eye. They are good optical analogs, which would have been entirely unanticipated on the basis of their weak homologous relationship. In another example, the Australian dingo (a placental mammal) and the Tasmanian wolf (a marsupial) are good ecological analogs, although the latter would be a very poor model if one were studying typical late embryological development in mammals. In any case, because modelling is based on relationships between organisms, it necessarily has reciprocal features: if A is a model of B, then B is a model of A.

In biomedical research, model selection generally begins with a search for close homologs that were judged at the outset of the research to be good analogs. Thus the spontaneously diabetic Wistar BB rat (discussed in the workshop on disease and aging, which is summarized in Appendix E) turns out to be an excellent model in the study of juvenile-onset diabetes, because the rat is a relatively close homolog of the human in terms of organ function, and between the ages of 60 and 120 days this strain develops diabetes with pathological characteristics almost identical to those of humans with the disease. Although other animals may be more closely homologous to humans, diabetic strains are not available and, therefore, cannot serve as models for this disease in humans. The use of Watanabe rabbits in the study of atherosclerosis and a strain of New Zealand black mice in the study of lupus erythematosus are additional examples of this principle. The search for animals with the same clinical manifestation of a disease as humans is an obvious route to the identification of animal models for biomedical research. Such systems seem so obviously useful as to require little further justification, but for purposes of completeness we elaborate on one such case.

For many years there were no adequate models to study human familial hypercholesterolemia and the attendant arteriosclerosis. Human skin fibroblast cells in culture possess a specific receptor for low density lipoprotein (LDL), which is lacking in cells from subjects with homozygous familial hypercholesterolemia. Thus, although certain biochemical aspects of the disease were studied in cell culture, no satisfactory animal models were available prior to 1975 to study the clinical aspects of the genetic defect. Then Kondo and Watanabe (1975) reported on a hyperlipidemic rabbit, which had been a spontaneous mutant. A homozygous strain that was subsequently bred has become known as the WHHL (Watanabe-heritable-hyperlipidemic) rabbit. The WHHL rabbits have been shown by analogy to be extraordinarily good models of humans with familial hypercholesterolemia insofar as the disease process is concerned. They have an LDL receptor deficiency in skin fibroblasts which the authors suggest will be a powerful tool for finding a significant role of LDL receptor-deficiency in the occurrence of the clinical syndrome of hyperbetalipoproteinemia (Tanzawa *et al.*, 1980).

By virtue of having a genetic lesion similar to that in humans with hypercholesterolemia, the WHHL rabbit is a fairly close model by homology of a circulatory system disorder. Studies over the past few years have shown many striking analogs between the disease process in WHHL rabbits and afflicted humans. Buja *et al.* (1983) described recent examples of modelling with the WHHL rabbit.

Cultured human skin fibroblast cells are models by homology (they possess the same genome and express some but not all of the same genes), and in the domains of biochemistry and cell physiology, they have proven to be good analogs for lipid binding. Thus, the WHHL rabbit and human cell cultures are providing excellent models for the study of a major disease and illustrate how one-to-one modelling can be extremely important.

In the process of developing the concepts of this report, it has been necessary to define two types of surrogate modelling, described as one-to-one and many-to-many.

One-to-one modelling. If in the study of a normal or pathological process or structure we find analogous behavior with respect to several features in two groups of organisms and no negative features, the organisms are models for each other with respect to those processes or structures, and studies of one are considered to have a high probability of yielding useful information about the other. For example, in a disease state of humans, if we can locate another organism that has the same range of symptomatic behavior, we are encouraged to use that organism as a model for studying the etiology, pathology, and therapy of the disease in humans.

Many-to-many modelling. If we have some process or state in an organism of interest and analyze it from a reductionist viewpoint into component features at several hierarchical levels, e.g., system, organ, tissue, cellular, or subcellular levels, we may then at each level note all the taxa in which analogous features appear. Each of these species is a model for the other with respect to those features. This is many-to-many modelling, the first *many* referring to the many features at various hierarchical levels that have emerged from the analysis and the second *many* referring to the many taxa at each level in which the features appear.

The usefulness of one-to-one modelling is illustrated by Brinkhous and Bowie (1977), who reviewed studies on the pigeon, dog, rabbit, pig, and nonhuman primates. They have shown how difficult it has been to develop good models of atherosclerosis, although pigs with von Willebrand's disease have been valuable in the study of spontaneous atherosclerosis. The search clearly relies on one-to-one modelling. Further exposition of the general approach can be found in a publication of the Department of Health, Education, and Welfare (1976).

One-to-one modelling describes the view of models that dominated the committee's initial discussions. The workshops and their subsequent analyses led to the adoption of many-to-many modelling -- a more general view of models that is introduced in the following paragraphs and developed in Chapter 5.

Investigators studying a phenomenon may analyze its various components at the organ, tissue, cellular, or subcellular levels and seek models for its different parts from the entire corpus of biological knowledge. This then allows them to study one organism or system in terms of related features from a variety of other organisms and other systems. In the new kind of epistemic structure that emerges, the matrix of biological knowledge replaces the one-to-one model as a source of analogs for reaching an understanding of problems. Within the matrix, homology and analogy still exist, analogy arising out of common strategies and common functional groups, and all the analogical behavior derived by using the same hardware and the same physical and chemical laws to solve similar problems or to perform similar functions.

A classical biological model (of the one-to-one type) can now be seen as one relationship within the larger context. If we are investigating some phenomenon in organism A, which has a certain relationship to the overall matrix, and in organism B, which has the same or very similar relationship to this matrix, then B is a good model for A for that specific study. Furthermore, experiments on B are likely to produce results that will be of great assistance in understanding A. Therefore, the modelling relationships are reciprocal. The associations need to be relevant only to the problem under study and need not be more general.

The body of biological knowledge is beginning to form a coherent and interrelated structure, but it lacks the tight theoretical formulation of physical science. As noted by Baldwin (1938) at the biochemical level, a general biology is emerging from our understanding of the vast number of interrelationships and common features that arose through organic evolution.

In a recent paper describing comparative neuroscience and its potential for advancing knowledge, Bullock (1984) makes a strong case for comparative studies and, thereby, for the use of models to provide data points in a complex matrix. The arguments he put forward for comparative neuroscience have been made for various aspects of the endocrine system (Roth *et al.*, 1983) and for other phenomena, other organisms, and other areas of biology. Because of its particularly cogent presentation, the summary of Bullock's paper is quoted here in its entirety:

> The brain has diversified and advanced in evolution more than any other organ; the variety of nervous systems and behaviors among animal species is thus available for our exploitation. Comparative neuroscience is likely to reach insights so novel as to constitute revolutions in understanding the structure, functions, ontogeny, and evolution of nervous systems. This promise requires pursuit on a wide front, in respect to disciplines and in respect to the species, stages and states compared. It also requires deliberate concentration on the differences among animals, in addition to the prevailing concern for the basic and common. Neglect of these challenges would be costly. Without due consideration of the neural and behavioral correlates of differences between higher taxa and between closely related families, species, sexes, and stages, we cannot expect to understand our nervous systems or ourselves (Bullock, 1984, p. 473).

The case for comparative studies in biomedical research is well made in the report of a study on research needs in endocrinology and metabolic diseases (National Institutes of Health, 1981). The report, compiled from the work of 18 task forces, including a committee on comparative endocrinology, contains the following statement:

> When viewed superficially, studies of comparative biology may seem esoteric or irrelevant to the human condition. The report of the task force on comparative endocrinology provides numerous examples which attest strongly to the contrary (National Institutes of Health, 1981, p. 42).

This assertion regarding comparative endocrinology can be applied to comparative studies in other biomedical research areas as well.

The committee's workshops have led inexorably to the conclusion that a theoretical biology or, to use Claude Bernard's phrase, a "theoretical medicine" is beginning to exist (Bernard, 1865). It is different from theoretical physics, which consists of a small number of postulates and the procedures and apparatus for deriving predictions from those postulates. But it is far more than just a collection of experimental observations. The vast array of information gains coherence when organized into a conceptual matrix through empirical generalizations and reductionist laws -- a construct that permits a view of models far more comprehensive than the committee envisioned at the outset of the study. This view is reflected in the concept of many-to-many modelling.

REFERENCES

Achinstein, P. 1968. Concepts of Science: A Philosophical Analysis. The Johns Hopkins Press, Baltimore, Maryland. 266 pp.

Baldwin, E. 1938. An Introduction to Comparative Biochemistry, 4th ed. Cambridge University Press, Cambridge, England. 179 pp.

Balls, M., R. J. Riddell, and A. N. Worden. 1983. Animals and Alternatives in Toxicity Testing. Academic Press, New York. 550 pp.

Beament, J. W. L., ed. 1960. Models and Analogues in Biology. Symposium of the Society for Experimental Biology, No. XIV. Academic Press, New York. 255 pp.

Bernard, C. 1865. Introduction à l'Étude de la Médecine Expérimentale. J.-B. Baillière & Fils. Paris. 400 pp. (An Introduction to the Study of Experimental Medicine. 1927. H. C. Greene, transl. Macmillan, New York. 226 pp.)

Brinkhous, K. M., and E. J. W. Bowie. 1977. Animal models of atherosclerosis involving the thrombotic process. Pp. 385-496 in Workshop on Thrombotic Process in Atherogenesis. Plenum Press, New York.

Buja, L. M., T. Kita, J. L. Goldstein, Y. Watanabe, and M. S. Brown. 1983. Cellular pathology of progressive atherosclerosis in the WHHL rabbit: An animal model of familial hypercholesterolemia. Arteriosclerosis 3:87-101.

Bullock, T. H. 1984. Comparative neuroscience holds promise for quiet revolutions. Science 225:473-478.

Department of Health, Education, and Welfare. 1976. Animal Models of Thrombosis and Hemorrhagic Disease. DHEW Publication No. (NIH) 76-982. U.S. Department of Health, Education, and Welfare, Washington, D.C. 203 pp.

Kaiser, H. E. 1980. Species-Specific Potential of Invertebrates for Toxicological Research. University Park Press, Baltimore, Maryland. 224 pp.

Kant, I. 1790. The Critique of Judgement. (J.C. Meredith, transl., 1952) Oxford University Press, Oxford, England. 434 pp.

Kondo, J., and Y. Watanabe. 1975. A heritable hyperlipemic rabbit. Exp. Anim. (Tokyo) 24:89-94.

Krebs, H. A. 1975. The August Krogh Principle: "For many problems there is animal on which it can be most conveniently studied." J. Exp. Zool. 194:221-226.

Krebs, H. A., and J. R. Krebs. 1980. The "August Krogh Principle." Comp. Biochem. Physiol. 67B:379-380.

Krogh, A. 1929. The progress of physiology. Am. J. Physiol. 90:243-251.

Margenau, H. 1977. The Nature of Physical Reality: A Philosophy of Modern Physics. Ox Bow Press, Woodbridge, Conn. 467 pp.

Maser, J. D., and M. E. D. Seligman. 1977. Psychopathology: Experimental Models. W. H. Freeman, San Francisco, California. 474 pp.

Mitchell, D. H., and T. E. Johnson. 1984. Invertebrate Models in Aging Research. CRC Press, Boca Raton, Florida. 195 pp.

National Institutes of Health. 1981. Proceedings of the Symposium on Trends in Bioassay Methodology: *In Vivo*, *in Vitro* and Mathematical Approaches. NIH Publication No. 82-2382. U.S. National Institutes of Health, Bethesda, Maryland. 371 pp.

National Research Council. 1977. The Future of Animals, Cells, Models, and Systems in Research, Development, Education, and Testing. National Academy of Sciences, Washington, D.C. 341 pp.

National Research Council. 1981a. Mammalian Models for Research on Aging. National Academy Press, Washington, D.C. 587 pp.

National Research Council. 1981b. Marine Invertebrates. Laboratory Animal Management Series. National Academy Press, Washington, D.C. 382 pp.

Ransom, R. 1981. Computers and Embryos: Models in Developmental Biology. John Wiley and Sons, New York. 212 pp.

Ross, D. M. 1981. Illusion and reality in comparative physiology. Can. J. Zool. 59:2151-2158.

Roth, J., D. Le Roith, J. Shiloach, and C. Rubinovitz. 1983. Intercellular communication: An attempt at a unifying hypothesis. Clin. Res. 31:354-363.

Russell, W. M. S., and R. L. Birch. 1959. The Principles of Humane Experimental Technique. Methuen, London, England. 238 pp.

Tanzawa, K., Y. Shimada, M. Kuroda, Y. Tsujita, M. Arai, and Y. Watanabe. 1980. WHHL rabbit: A low density lipoprotein receptor-deficient animal model for familial hypercholesterolemia. FEBS Lett. 118:81-84.

Thompson, D. W. 1917. On Growth and Form. Cambridge University Press, Cambridge, England. 793 pp.

Turing, A. M. 1936. On computable numbers, with an application to the Entscheidungs-problem. Proc. London Math. Soc. Ser. 2 42:230-265.

4

Evaluation of Models for Biomedical Research

> Unless we recognize our innate biases in animal model choice, we limit our potential as experimenters. Two biases seem common from my observations. First is the anthropomorphism that we all seem to get from the monkeys in zoos and circuses, coming as it does long before we aspire to be scientists. Second is for the animal or animals with which we worked during our early days in our fields. Both of these are easy to understand and are forgivable. What has neither of these saving attributes is an unwillingness to consider the entire biologic kingdom as a source of possible models of one or another human functions, normal or diseased (Prichard, 1976, p. 172).

The workshop topics were selected as case studies for sampling research areas with relevance to the mission of the National Institutes of Health (NIH). Other areas might also have been profitably sampled. Within each area selected, the workshop participants represented only a few of the investigators studying designated problems in significant, novel ways. It is therefore essential for the reader to refrain from overinterpreting the comments made about any particular system. Although those discussed are often exemplary, others, either not discussed or perhaps yet to be discovered, might be superior.

As explained in Chapter 2, the committee responded to its broad charge initially by selecting for analysis five broad, basic biomedical research areas. During the workshops, presentations were made by experts who used a variety of experimental materials. The first of these workshops was focused on cellular immunology, which of necessity is investigated with *in vitro* or *in vivo* vertebrate systems. The next four workshops on learning, regulation, diseases and aging, and development were more broadly based taxonomically, and in each an effort was made to identify authorities on a variety of models. Unlike the others, the sixth workshop had a particular type of modelling (mathematical) as its common denominator and different types of biological problems as its range of focus.

Through the workshops, the committee gained valuable information on the extent to which studies on diverse organisms provide new insights in biomedical research. It also became evident how different model systems could be applied to different aspects of the same problem. The learning workshop, for example, illustrated how investigation of a biological process progresses through the use of models at different levels of biological organization as well as different phylogenetic (evolutionary) status (Appendix C).

Workshop participants facilitated comparisons across taxa and between different hierarchical levels. Learning, for example, was modeled in bacteria, invertebrates, lower vertebrates, and primates, as well as through mathematical modelling and in tissue preparations *in vitro*.

By making comparisons and identifying some opportunities for comparisons that arose during the workshops, the committee was able to draw some conclusions regarding the roles of different types of models across the broad spectrum of biological research. In a sense, it was able to organize knowledge about biological systems and processes as a function of level of biological organization. The committee explored the generalizability of model behavior as a function of phylogenetic position, which correlates roughly with the level of complexity. In doing this, it engaged in many-to-many modelling, described in greater detail in Chapter 5. The following may be viewed as an attempt to define some of the parameters of the matrix of biological knowledge, where the matrix is the second "many" of many-to-many modelling.

HISTORICAL PERSPECTIVE

From the earliest days of modern biology and medicine, animals have been widely used to gain insights into human biology as well as general biology. Animal dissections and experiments marked the growth of biology and medicine and became central to their progress. Because ethical considerations generally preclude research on humans, surrogates are essential to every aspect of biomedical research.

Understandably, the similarities in structure and function among mammals made them obvious candidates for research. As experimental biology and medicine matured in the last century, rodents -- rats, mice, guinea pigs, and hamsters -- came into favor because their small size made them suitable for laboratory experimentation and because they could be bred with ease in captivity. Less well appreciated has been the ongoing role of lower vertebrates, invertebrates, and microorganisms in biomedical research.

The importance of various organisms to progress in biomedical research during this century can be seen in Appendix A, which lists the organisms used in research that has led to Nobel Prizes in physiology or medicine. This compilation demonstrates the conspicuous role played by nonmammalian organisms in research that is generally believed to be of the highest caliber, the most enduring influence, and the most importance to biomedical science.

Some organisms are studied as causative agents of disease (e.g., studies on the mosquito's role in the transmission of malaria resulted in an award to Ronald Ross in 1902). In most cases, however, the experimental organism served as a model in one-to-one or many-to-many modelling. Golgi and Ramon y Cajal shared the 1906 award for elucidating aspects of neuroanatomy through their studies of birds and reptiles as well as dogs, horses, and humans. In 1924, Hill and Meyerhof were recognized for their studies on muscle physiology, in which preparations derived from frogs were used. Loewi, honored in 1936 for his insights into the chemical transmission of nerve impulses, also used frogs. Other Nobel laureates have used invertebrates to study nerve function (Hodgkin and Huxley in 1963) and visual processes (Hartline and Wald in 1967). The use of yeast, bacteria, and viruses to study the molecular basis of genetics has resulted in several Nobel citations.

Higher plants have also been important as sources of model systems. Corn (*Zea mays*) was the system used by McClintock in the discovery of mobile genetic elements, which earned a Nobel award in 1983.

The survey of an international series of awards is but one indication of the contributions made by nonmammalian organisms to biomedical research. Some roles played by nonmammalian systems in developmental biology and neurobiology are shown in the following examples:

- Discovery and analysis of meiosis: ascaris, annelids, sea urchins, amphibians.

- Elucidation of the events of fertilization (e.g., cell fusion, acrosome reactions, cortical granule exocytosis, and pronuclear pairing): sea urchins, starfish, amphibians.

- First analysis of developmental stages in anatomical terms; origins of descriptive embryology; first understandings of organogenesis: amphibians, birds, pigs, mice.

- Early studies of the action of sex steroids in sexual dimorphism: amphibians, birds.

- First construction of vertebrate fate maps and recognition of germ layers; early concepts of cell fate, potency, and determination: amphibians, birds.

- Discovery of primary induction (the induction of the neural plate by chordamesoderm) and of induction of the lens by the optic vesicle: amphibians, birds.

- Recognition of the central role of epithelial-mesenchymal interactions in the development of ectodermal and endodermal organs: amphibians, birds, mice.

- Discovery of the role of the neural crest in the development of, for example, the autonomic nervous system, head structures, adrenal medulla, pigment cells: birds.

- Nuclear equivalence demonstrated by nuclear transplantation: amphibians.

- Discovery of cell sorting by differential adhesion: amphibians, birds, sea urchins.

- Discovery of transcellular ion fluxes: algae, marine invertebrates.

- First experimental teratogenesis to produce spina bifida, microcephaly, and Siamese twinning: fish, amphibians, birds.

- Basic studies of the cell cycle, mitosis, and cytokinesis (e.g., spindle isolation): yeast, sea urchins, starfish, surf clams, amphibians.

- Analysis of DNA base sequences needed for tissue-specific, stage-specific, and hormone-specific gene expression: fruit flies.

- Discovery of gene amplification: amphibian oocytes.

- Cytoplasmic localization of determinants: ascidians, other marine invertebrates.

- First apparent support for the concept that "ontogeny recapitulates phylogeny": comparative embyology of fish, amphibians, birds, mammals.

- First modification of genotype by DNA transformation: bacteria, yeast, fruit flies, mice.

- Ionic basis of membrane excitability in nerve and muscle cells: squid stellate ganglia.

- Physiological mechanism of neurotransmitter release at chemical synapses: squid stellate ganglia, frogs.

- Demonstration that GABA (gamma-aminobutyric acid) is a neurotransmitter and elucidation of ionic mechanisms underlying its inhibitory synaptic effect: crustacean inhibitory neuromuscular junction.

- Glutamate as a synaptic neurotransmitter: crustacean and insect excitatory neuromuscular junctions.

• Establishment, refinement, and elucidation of the concept of neuromodulation and demonstration that biogenic monoamines such as octopamine and serotonin can be neuromodulators: crustacean and insect preparations.

• Establishment of histamine as a neurotransmitter: molluscs.

• Discovery of many principles of neurosecretion: insect central nervous systems (CNS).

• Physiology, transduction mechanisms, and sensory coding mechanisms of chemosensory (olfactory and gustatory) receptor cells: insect antennal, maxillary, and tarsal chemoreceptors.

• Discovery of pheromones and elucidation of mechanisms of pheromonal signalling, detection, and control of behavior: insects.

• Advances in understanding mechanosensory feedback control of movement and central mechanisms of motor control: crustacean and insect preparations.

• Synaptic and modulatory mechanisms underlying simple forms of learning and behavioral plasticity: molluscan preparations.

• Physiology of glial cell membranes and the functions of glia: annelids (leech).

• Precise and complete analysis of cellular lineages giving rise to neurons in developing CNS: nematodes (*Caenorhabditis*), annelids (leech), insects (locust).

• Mechanisms guiding axonal growth cones during development of the CNS: insects.

• Specificity of cell-cell interactions and precision of synapse formation during development and regeneration of the nervous system: insects.

• Demonstration of molecular differentiation of different types of neurons and establishment of monoclonal antibody techniques for studying neural development and connectivity: insects, leeches.

• Comparison of genetic and epigenetic contributions to neural development: flies, nematodes.

• Cellular interactions and physiological mechanisms underlying the blood-brain barrier: insect CNS.

The oocyte of the African toad *Xenopus* has been extraordinarily valuable because of the ease with which macromolecules can be injected into it. Many studies have exploited this property to learn among other things about RNA processing, chromatin structure, and nuclear membrane formation. The oncogene homologs found in yeast are examples of non-mammalian systems that should provide important information relevant to mammals.

An important process in development concerns assembly and disassembly of macromolecular aggregates such as the mitotic and meiotic apparatus, the cytoskeleton, and other cellular scaffolding systems of known and unknown functions. The most sophisticated studies on macromolecular assembly come from phage morphogenesis. Principles of building, such as those governing the development of macromolecular structures, have been learned in studies of the assembly of viruses, especially bacterial viruses.

The following sections of this chapter contain discussions of each potential source of biomedical research models for which the NIH requested evaluation. These evaluations were greatly facilitated by the contributions of workshop participants, which are summarized in Appendices B through G.

MICROORGANISMS

In the workshops, microorganisms such as procaryotes, protozoa, and other protists received less attention than did higher eucaryotes. This was partly because of the topics selected for the workshops by the committee members, who realized that microorganisms have become so widely accepted as models in metabolism, genetics, and biochemistry that they required no special attention. The microorganisms discussed in this chapter are of special importance because they can be used to study relationships that go beyond modelling at the lower hierarchical levels, thereby expanding the previously expected possibilities for this broad group of organisms.

Microorganisms were a major theme only in the workshop on development (Appendix F). Walker spoke about the use of the procaroytic *Escherichia coli* to study regulation of gene expression, which is clearly a component of developmental biology. The induction or inhibition of gene products can be studied at the chemical level in bacteria. Herskowitz discussed determination of mating types in the yeast *Saccharomyces cerevisiae*, an example of the study of gene expression in a sexually reproducing unicellular eucaryote. Trypanosomes are protozoans with genetic coding capable of specifying hundreds of surface glycoproteins, only one of which is expressed at a time. Discussed by Agabian, this is a developmental control of gene expression, but in an even more complex organism. Each of the different microorganisms is being studied for specific purposes, ranging from basic knowledge to disease control, but there is every reason to expect that fundamental mechanisms of gene expression will have broad intertaxonomic features and that information emerging from studies of trypanosomes will be applicable to the study of normal and pathological development of human embryos.

The study of the slime mold *Dictyostelium*, described by O'Dell in the workshop on mathematical models (Appendix G), is an example involving unicellular microorganisms in the ameboid stage of the life cycle. O'Dell demonstrated not only a mathematical model for studying the behavior of each cell in relation to the integrated behavior of a cell aggregate, the pseudoplasmodium, but also a method to study cell-cell interactions, a feature of morphogenesis that occurs in higher forms.

In the learning workshop (Appendix C), Berg discussed the concept of short-term memory in bacterial chemotaxis. Procaryotes and other organisms lacking neurons and neural junctions clearly do not provide appropriate models for learning in higher organisms. Nevertheless, the molecular mechanisms used by bacteria to sense and respond to the chemical environment and its rate of change are of interest in research on higher organisms, because they may provide clues to the mechanisms controlling some memory components of learning in more complex organisms.

In the regulation workshop (Appendix D), Roth discussed the finding that some peptides in microorganisms have amino acid sequences similar to vertebrate neuropeptides. These studies indicate the reciprocal nature of models and of biological information transfer. Knowledge of the role of these molecules in higher organisms suggests experiments to test their possible role as signaling agents in bacterial cells. Here higher vertebrates are in a sense serving as models for microoganisms, or at least for certain processes in microoganisms. The implications of many-to-many modelling to the matrix, both theoretical and applied, are illustrated at another early phylogenetic level, where it has been shown recently that yeast have receptors for estrogen that appear identical in terms of affinity to those from the rat uterus.

Thus, there is also a possibility of reciprocal modelling between yeast and rats. These newer findings, spanning such a broad taxonomic range, provide further evidence of the extent to which all of biology is interelated and modelling is an exploration of these interconnections.

INVERTEBRATES

There are more than 2×10^6 species of insects. Species of beetles alone outnumber species of all other animals and plants combined. Many scientists believe that invertebrates, and insects in particular, should be studied for their own sake, apart from the fact that they are of economic and medical importance and occur in such vast numbers. Such investigations are initiated by an intrinsic interest in comparative invertebrate biology and biochemistry but often lead to results with more general applicability. Thus, Keilin's interest in the fate of larval hemoglobin in the botfly (*Gastrophilus*) resulted in his discovery of the cytochromes (Keilin, 1925); Fraenkel's investigations of the factors essential for the growth of the mealworm (*Tenebrio*) larva culminated in identification of a new cofactor, L-carnitine (vitamin B_T) (Fraenkel, 1948); and Roth and Porter's ultrastructural analysis of yolk uptake by the oocytes of the mosquito *Aedes* provided the first description of coated vesicles and their role in cellular uptake (Roth and Porter, 1964).

For the biochemist, invertebrates are of interest because of some unusual reactions or products, or because a well-known biochemical reaction is highly exaggerated. For example, Sir Gowland Hopkins' interest in the nature of the white pigment observed on the wings of a butterfly led to the structure and biosynthetic pathway of a new class of compounds, the pteridines, and ultimately to the vitamin folic acid (Hopkins, 1895). More recent studies in Kreil's laboratory on a polypeptide from bee venom may lead to similar generalizations in the future (Suchanek *et al.*, 1978).

At the molecular level, there is little question that the fruit fly (*Drosophila*) with its well-understood genetics, numerous mutants, transposons, and polytene chromosomes will be an organism of choice in molecular biology. This is fitting, since it was in part the work on *Drosophila* eye pigmentation by Beadle and Ephrussi that led to the hypothesis that each gene controlled a single enzyme -- a concept that was the foundation block of molecular genetics, the cornerstone of modern molecular biology (Ephrussi, 1942).

Regulation

The contributions of invertebrates to research on regulation have been diverse and important. One has only to recall that Hodgkin's and Huxley's Nobel Prize-winning studies of the squid giant axon provided the basis for the concept of the ionic nature of the action potential (Hodgkin and Huxley, 1952; Appendix A) and that much of our knowledge of the physiological mechanism of neurotransmitter release stems from the work of Katz and Miledi on the squid stellate ganglion (Katz, 1969).

The insect brain was the first nervous tissue identified as having endocrinological properties, and the concept of neurosecretion owes much to the study of invertebrates by Scharrer (see Appendix D). Thus it was not surprising that the first *in vitro* reprogramming of a brain-centered photoperiodic clock was demonstrated in an insect system (Bowen *et al.*, 1984). These and other neuroendocrinological advances have been possible since invertebrates display the following facilitating characteristics, among others: the absence of classical immune and circulatory systems, allowing transplantation of organs between organisms and between species; responses to environmental cues characterized by overt physiological changes; short life cycles, permitting investigators to conduct a multitude of experiments under carefully controlled conditions; and relatively simple nervous systems.

In the last few years, there have been remarkable advances in invertebrate neurobiology, including fate mapping of the entire nervous system of a nematode (see Appendix F and section on Development in this chapter), cell recognition during neuronal development, and neuropeptide control of behavior. The technology developed for invertebrates in the latter two areas will probably be adapted for analogous studies on more complicated vertebrates.

Goodman *et al.* (1984) have utilized the relatively simple nervous system of insect embryos to analyze the cellular and molecular basis of cell recognition during neuronal development. Some of their earlier studies were carried out with grasshopper embryos, because the cells are easily accessible. More recently, they have used *Drosophila* because its genetics are so well known. These investigators have shown that the growth cones of specific neurons express selective affinities for specific axonal surfaces and that this is the basis for stereotypic fasciculation patterns of axon bundles. The results of their studies indicate that very early in development, cell lineage and cell-cell interactions lead to the differential expression of cell recognition molecules on the surface of groups of embryonic neurons whose axons then form a "bundle." They have prepared monoclonal antibodies to these putative surface molecules. With the available genetic probes for *Drosophila*, the genes for these specific cell surface molecules may be cloned and their function determined. Ultimately, the result may be an understanding of how a nervous system is "wired" during development -- one of the major unanswered questions in biology.

Scheller *et al.* (1984) have been investigating the molecular biology of the neuroendocrine system of the mollusc *Aplysia*, which was chosen because of the relatively small number but large size of the neurons constituting its nervous system. Using recombinant DNA methodology, they have isolated the genes that encode the peptide precursors found in specific neurons already known to have a particular function in the reproductive behavior of the animal. Some of these peptides have multiple biological activities and are expressed in neurons that also synthesize classical transmitters. These investigations have already provided the basis for unraveling the molecular mechanisms that govern simple behavioral phenomena. Although the *Aplysia* system is considered to be relatively simple, Scheller *et al.* (1984) estimate that up to 40 different neuropeptide precursors and more than 200 biologically active peptides may perform neuroendocrine functions in this primitive organism. If the latter is correct, one can only look with pessimism at the prospects for any immediate success along the same lines with the mammalian nervous system, which might contain an even greater number of biologically active peptides that control the more complex behavioral patterns of these organisms. This work illustrates the utility of invertebrate models in studies to gain fundamental information on the workings of the nervous and neuroendocrinological systems.

Aside from the successful pursuit of basic knowledge, there have been rewarding serendipitous results. For example, Benzer has been involved for many years in studies of the basis of *Drosophila* behavior and recently generated 148 monoclonal antibodies directed against *Drosophila* neural antigens (Fujita *et al.*, 1982). Although developed for studying correlations between specific neurons and behavioral events, the antibodies were tested against human brains and half of them reacted with one or several sites in the human central nervous system (Miller and Benzer, 1983). Regional, cell class, and subcellular antigens were identified, and some of the antibodies recognized neuronal, glial, or axonal subsets. Furthermore, some antibodies reacted with similar antigen patterns in both organisms. The authors reasoned, "In various neurologic diseases, there is selective vulnerability of certain neurons or glial cells. For example, in amyotrophic lateral sclerosis, motor neurons degenerate; in Huntington's disease, cells of the caudate nucleus deteriorate; in some cases of Alzheimer's disease, cells in certain basal forebrain nuclei are lost. The *Drosophila* monoclonal antibodies MAbs may be useful as tags for sorting specific cell types to identify the molecular profiles of such selectively vulnerable cells and to detect missing or novel antigens in diseased tissue." Moreover, they reported, "Many human neurological defects are hereditary, but progress in human disease has been stymied often by the lack of model systems. *Drosophila* mutants also display hereditary pathologies such as brain degeneration in the *drop dead* mutant and muscle defects resembling nemaline myopathy in the *wings-up* mutant. With currently available recombinant DNA technology, it may well be feasible to transfer a selected gene from human to fly in order to study its function" (Miller and Benzer, 1983).

The importance of invertebrates as model systems is underlined by the remarkable findings in the last several years that virtually every peptide and steroid hormone known in humans is immunochemically similar to substances found in insects and other invertebrates. These include insulin, glucagon, somatostatin, gastrins, substance P, vasopressin, neurophysin, enkephalin, endorphin, adrenocorticotropin (ACTH), calcitonin, thyroid-stimulating hormone (TSH), luteinizing hormone-releasing hormone (LHRH), prostaglandins, estrogens, and androgens (see Kramer, in press). The functions of these substances may, of course, be different in humans and in invertebrates. In addition, insects utilize their own unique peptide and steroid hormones as well as a sesquiterpene hormone, the juvenile hormone (see Appendix D).

The work of Roth and colleagues on unicellular eucaryotes revealed the presence of an insulin-like molecule in fungi and protozoa, which suggested that the molecular origins of insulin go back as far as the earliest eucaryotes (Le Roith *et al.*, 1980) and that the molecule may act as a signal, i.e., intercellular or intracellular messenger, in these as it does in mammals. Further investigations have identified in procaryotes and in unicellular eucaryotes peptides such as thyrotropin, somatostatin, ACTH, endorphin, and even a "prohormone" containing ACTH and endorphin-like immunoreactivity on the same molecule as in mammals (Le Roith *et al.*, 1983).

Steroid hormones appear to be old molecules phylogenetically (Tata, 1984). It has been known for many years that vertebrate sex hormones as well as corticosterone are present in invertebrates, although like the peptides noted above, their function is unknown; that the water mold *Achlya* utilizes a steroid hormone (a pheromone) for initiating and coordinating the sexual process; and that the action of the antheridiols closely resembles the effect of estrogen and progesterone on the rat uterus and chick oviduct (Timberlake and Orr, 1984).

Also of interest here is the work of Feldman's group, demonstrating that yeasts can synthesize estradiol and probably have a classic estrogen receptor system (Loose *et al.*, 1983). Indeed, these investigators suggest that the presence of estrogens in women inhibits the transformation of mycelium to yeast, which could explain why women are more resistant to the yeast infection paracoccidioidomycosis than are estrogenless men.

This all infers that certain deoxyribonucleotide sequences have been preserved through billions of years of evolution. In some cases, this has been proved either by sequencing the homeobox (specific gene sequence) of *Drosophila* and humans or by analyzing the amino acids of the head-activating factor of *Hydra* and humans (Schaller and Bodenmuller, 1984). The implication of these findings for biomedical research is that there are unifying principles at the cellular and molecular level for the regulation of cellular activities and that these invertebrates contain molecules of potential interest in the investigation of regulatory processes in humans. More and more evidence is accumulating to support this view.

Development

Tomes could be written about the contributions of invertebrates to developmental biology, since many of the key principles guiding this field were formulated in the late 19th and early 20th centuries from work on the embryos of echinoderms, annelids, and amphibians. Continuing in modern times, the concept of long-lived, masked messenger ribonucleic acid (mRNA) was derived from studies of the sea urchin embryo (Raff *et al.*, 1972). The field has been revolutionized recently by the discovery of transposable elements and their ability to insert genes into the germ line of *Drosophila* embryos (Rubin and Spradling, 1982; Spradling and Rubin, 1982). This provides the possibility of analyzing not only *Drosophila* development at the molecular level but also perhaps the regulation of foreign genes inserted into *Drosophila*.

The nematode *Caenorhabditis elegans*, discussed in Appendix F, is another organism that has been used for morphogenetic analysis. This system was developed by Brenner and Sulston at Cambridge. As noted previously, the development of its entire nervous system has been mapped and various cell lineages and mutants, including homeotic strains, have been described (Sulston *et al.*, 1983). These findings have been used to study the development of the gonad, for which lineages have been mapped and a large number of *lin* mutants identified. Thus, *C. elegans* may become an ideal model for studies of cell determination at the molecular level.

Ascidians, so-called invertebrate vertebrates, produce eggs with pigmented regions that allow the visualization of cell determinants. In recent studies, Jeffery *et al*. (1984) have attempted to learn whether cell determinants may be composed in part of maternal mRNA, how these determinants are localized in the embryo, and how the determinants are assayed (Appendix F). Their approach cannot make use of genetic analysis, but promises important answers to these fundamental questions in developmental biology. There are many such "ideal" systems among invertebrates, but it is likely that the most definitive research will once again be conducted on *Drosophila*, because its genetics are so well understood (see Appendix F) and some of its critical gene sequences are shared with higher vertebrates.

A DNA sequence of *Drosophila* associated with homeotic genes (mutant genes that cause the development of a body part in place of the normal part, e.g., a leg instead of an antenna; see Bender *et al*., 1983) has now been identified in representatives of several other phyla, including humans (Miller, 1984). It is likely that this sequence occurs in genes that control developmental processes in many organisms. It encodes about 60 amino acids and has only been found in segmented organisms. The most recent speculation is that the resulting protein binds to DNA and regulates a variety of developmental genes. McGinnis *et al*. (1984) found that the homeo protein closely resembles the Alpha 1 and 2 mating proteins of yeast that control the expression of yeast mating genes and genes that control cell differentiation. Since the basic skeletal organization, muscles, and nervous system of vertebrates arise from repeated embryonic structures, the somites, vertebrates can be considered segmented animals. The homeobox is absent from the genome of non-segmented organisms, including nematodes, sea urchins, and bacteria. It now appears that a homologous sequence is present in the African toad *Xenopus*. If it has the same function as in *Drosophila*, it would identify an important gene controlling normal development in vertebrates (Miller, 1984).

A continuing mystery in developmental biology is the chemical nature of morphogenetic agents, inducers, and other chemical messengers of development. There is a great deal of evidence that such messengers exist, but not one has been characterized structurally. It is therefore of interest that the hormones controlling the postembryonic development and metamorphosis of insects (ecdysteroids and juvenile hormones; see Appendix D) have been identified in insect embryos and that their titers can be correlated with specific morphogenetic events (Hoffmann and Lagueux, in press). If it can be shown that embryogenesis is analogous to postembryonic development in these organisms insofar as hormonal control is concerned, these hormones may prove to be the embryonic inducers that have been discussed over the past 6 decades.

Developmental abnormality has emerged in recent years as a major cause for concern in the health hazard evaluation of environmental chemicals and pharmaceuticals. In evaluating a variety of *in vitro* systems for their potential use in teratogenicity screening, Wilson (1978) calls attention to test systems that use intact nonmammalian embryos such as *Drosophila* larvae, sea urchins, fish, and chick embryos.

Learning and Behavior

Invertebrates have been of importance in research on learning and behavior (see Appendix C), and studies on invertebrates such as *Drosophila* and *Aplysia* will make major contributions in future years. Over the years, the pioneering work of Pittendrigh on *Drosophila* has been vital to our understanding of biological rhythms (Appendix D), and recently Bargiello and Young (1984) have isolated from that organism 90-kb pairs of DNA that appear to be responsible for some rhythmic behavior. Genetic analysis has led the authors to suggest that the DNA sequences required for rhythmic behavior may reside within a 7.1-kb DNA fragment. This lays the groundwork for a molecular analysis of biological rhythms.

The work of Gelperin and Culligan (1984) with an isolated lip-brain preparation derived from food-conditioned slugs, *Limax*, demonstrated that these preparations express *in vitro* the learning acquired *in vivo*. A number of molluscs, including *Aplysia*, have been extremely useful in analyses of associative learning (Appendix C). Long-lasting synaptic events with associative components have been described in detail (Kandel *et al*., 1983). This relatively "simple learning" may ultimately be described in molecular terms (Scheller *et al*., 1984). The studies of Benzer and his colleagues on the neural basis of behavioral mutants of *Drosophila* also hold great promise in this regard (Hotta and Benzer, 1972).

NONMAMMALIAN VERTEBRATES

It is difficult to envision how the biological sciences could have progressed to their present level of sophistication without the use of nonmammalian vertebrates, including fishes, amphibians, reptiles, and birds. They provide availability, convenient maintenance, both generalized and specialized functions, and relatively close relatedness to mammals. The examples given below are representative; they are far from exhaustive.

Although the squid giant axon was used by Hodgkin and Huxley (1952) in their classic analysis of action potential, the mechanism of propagation in vertebrate myelinated nerve is different. Excitable membrane is restricted to nodes of Ranvier, and excitation jumps from node to node by saltatory conduction. The techniques established for studies of myelinated nerve were ultimately combined with the Hodgkin-Huxley approach to demonstrate the action potential mechanism at nodes, which did turn out to be in most respects the same as that in the squid (Frankenhaeuser and Huxley, 1964).

Subsequent work on mammalian myelinated fibers showed a few further differences, but the frog provided an important if not essential first model system. Most of the basic properties of chemical transmission were derived from the study of the frog neuromuscular junction -- the synapse between nerve and skeletal muscle. The experiments of Loewi (1933) with the frog heart established that neural transmission could be mediated chemically. More detailed knowledge of the mechanism came with intracellular recording and the experiments of Katz and his colleagues (Katz, 1969), who demonstrated that a transmitter was released in small packages, or quanta. Shortly thereafter, these packages were given a morphological correlate in synaptic vesicles as new techniques allowed application of electron microscopy to biological tissues. These investigators demonstrated the requirement for calcium, and in correlated experiments on the squid synapse, showed that influx through calcium channels activated by the presynaptic impulse was an essential step in transmitter release. The neuromuscular junction also was used for determining the mechanism of transmitter action on the postsynaptic membrane. A localized permeability change that occurred differed from that produced by the impulse. Pharmacological manipulations, especially those involving microscopic application of transmitter, blockers, and esterase inhibitors, set a standard for the analysis of pharmacological actions and led to many clinically important uses. The isolation of the transmitter receptor can be traced directly to this work, and as described elsewhere in this report (Chapter 5; Appendix E), our knowledge of myasthenia gravis arises largely from studies on rats and from the use of a number of different nonmammalian vertebrates -- studies that addressed specific biological questions rather than the general understanding of a particular disease.

Fishes have figured prominently in studies leading to our understanding of diabetes mellitus. In some species the islet tissue, which is now known to be the source of insulin, comprises large nodules in the mesentery, completely separated from the exocrine pancreas (Epple, 1969). An extract of this tissue proved to cause hypoglycemia when injected in rabbits, thereby providing the first direct evidence that the islets were the source of the hormone (Macleod, 1922). The isolated islets of fishes are still being used in studies of hormone biosynthesis, cell excitability, and cell interaction.

Not long after the demonstration of chemical neurotransmission in the 1950s, many electrical synapses were found in nonmammalian vertebrates (Bennett, 1977). Knowledge of the morphological and physiological properties of the synapses in these species aided and promoted subsequent identification and characterization of these synapses in mammals. It is clear that both chemical and electrical modes occur in animals from flatworms to humans.

Much experimental material for the study of muscle contraction has come from frogs. The sliding filament model was formulated from electronmicroscopy of frog skeletal muscle (Huxley and Hanson, 1954), and the same muscles were used as the source of single fibers to study the relationship between the degree of overlapping of filaments and tension

(Gordon *et al.*, 1966). Modern x-ray analysis is revealing the changes in molecular structure that cause macrosopic movement (Huxley *et al.*, 1981).

Studies of nonmammalian vertebrates have contributed importantly to ethology. The concepts of fixed-action pattern, releasers, critical periods, and imprinting were derived largely from the study of birds and fishes. Learning of bird song, as described by Nottebohm (see Appendix C), is an annual process in many species. The new and unexpected finding is that old neurons die and new neurons are formed to replace them each year in the song-controlling nucleus. Forgetting and relearning appear to involve actual replacement of the mediating neurons.

The retina has many similarities throughout the vertebrate phylum. Various components are more highly developed in different species, sometimes divergently, but the general organization remains the same (Polyak, 1957). Our current knowledge has depended greatly on fish, amphibia, reptiles, and birds as well as on mammals. The original intracellular recordings showing absence of impulses and the unexpected hyperpolarizing responses of the photoreceptors were derived from studies on a salamander, although presaged by earlier studies in fishes (Dowling, 1979). Both fishes and turtles are now being used to study electrical interactions between horizontal cells (Piccolino *et al.*, 1984). The pigeon retina is constructed with precise lamination correlated with the presence of specific peptide neurotransmitters (Brecha *et al.*, 1984). The concept of sensory processing in the retina was largely formulated from studies on the frog (Lettvin *et al.*, 1959).

There are many similarities as well as differences in embryonic development throughout the vertebrate class. The fundamental process of gastrulation was first described in amphibia by Vogt in the 1920s, and eggs that develop externally remain important in studies of embryological development. Such fundamental questions as how polarity is determined are still being investigated in nonmammalian organisms, because similar processes in mammals are much less accessible (see Appendix F).

Vertebrate body plans are very similar to one another in early embryonic stages -- a finding leading to the phrase, ontogeny recapitulates phylogeny. Although now recognized as an oversimplification, we regard gill slits, branchial arches, cranial nerves, the oculomotor system, vestibular apparatus, and many other anatomical features as fundamentally similar throughout the vertebrates. Studies on amphibian embryos have been essential to our concepts of the competency of nuclei to form normal embryos in somatic cells (Gurdon and Laskey, 1970), and *in vitro* preparations of amphibian oocytes are important in evaluating the ability of mRNAs to direct the formation of specific proteins (Lane, 1981). The molecular biology of development in vertebrates relies heavily on these preparations, and the rapid bioassay provided by amphibian oocytes is likely to remain a standard technique in evaluation of gene expression. Features of bullfrog tadpoles that make them useful as biomedical research models are summarized in Appendix D.

CELL AND TISSUE CULTURE

Cell, tissue, and organ cultures derived from invertebrates and vertebrates have become such central elements of biological research that they are no longer viewed primarily as models but, rather, as a body of techniques widely exploited in biomedical research.

In the early 1900s, Harrison (1907) demonstrated the growth of nerve cell processes from spinal ganglia embedded in clotted lymph. These experiments were both the beginning of tissue culture and a dramatic reaffirmation of the cell as the fundamental, discrete, and potentially autonomous unit of life. It was soon realized by many that culture of cells and tissues *in vitro* offered a new and powerful approach for analyzing problems of animal physiology and embryonic development. One could now take apart the organism cell by cell, culture the cells in a controlled environment, and observe the effects of environmental variations on the course of development or physiological function. There were two major problems, however. First, differentiated cell lines were not developed until more than 50 years after Harrison's initial experiments (Levintow and Eagle, 1961). Second, serum or some biological fluid was an obligatory component of cell culture media, making it very difficult to identify the critical factors in the *in vitro* environment (Hayashi and Sato, 1976).

For several years Sato *et al*. (1970) were concerned with the establishment of cultured cell lines that retained the differentiated properties of the tissue of origin. This turned out to be a very useful experience for the investigators when they studied the role of serum in cell culture medium (Hayashi and Sato, 1976). Many people had tried to understand this role by analysis -- that is, by fractionating serum to isolate the active component (Fisher *et al*., 1958; Holley and Kiernan, 1974; Jacquez and Barry, 1951; Lipmann, 1971; Temin *et al*., 1972).

In retrospect, an analytical approach is awesomely difficult because of the complex synergistic interactions between factors and the minute quantities of hormones in blood. Many good scientists took the analytical approach probably because they had little confidence in cultures as faithful models of cells in the whole animal. Others concerned with the differentiated functions of cells in culture are confident that they reflect the physiological behavior of cells in the animal. To arrive at a resolution of this problem one must ask: Which of the animal's tissues is the source of the cultured cells? What are the normal stimuli for their differentiated function and proliferation in the animal? This naturally leads to the hypothesis that hormones must be among the key components of serum. Sato (1975) stressed that a role of serum in cell culture medium is to provide hormones. This role has since been expanded to include provision of basement membrane components (Barnes and Sato, 1980).

Hormonally defined media are now commonly used in cell culture. Researchers engaged in this activity have defined a new field -- cellular endocrinology, which has two main areas of activity -- hormone mechanisms and integrated physiology. Cell cultures lend themselves readily to mechanistic studies because of the technical ease with which they can be manipulated. The questions that are asked are so basic that the appropriateness of the model systems are never in doubt. Three new findings about the mechanism of hormone action have recently emerged from *in vitro* studies: the activation of proto-oncogenes by hormones, the hormonally induced production of diacyl glycerides and their activation of protein kinases, and the involvement of hormone-receptor complexes in the activation of nucleases (Cohen, in press; Kelley *et al.*, in press; Nishizuka, 1983). The extrapolation of culture results to whole animal physiology will be much more difficult.

The workshop on model systems in cellular immunology (Appendix B) dealt with *in vivo* and cell culture studies, moving back and forth between the two. Cloning of individual cell types and the study of mixed clones represents an intermediate step between cells and organisms. The final test of culture methods will probably come from two sources. First, factors discovered in culture will probably be implicated in human disease. No deficiency diseases have yet been discovered in tissue culture, but it is not far-fetched to imagine that an important disease like acquired immune deficiency syndrome (AIDS) might some day be traced to a deficiency of some lymphokine. The second source of legitimatization should be theoretical. If elucidation of the interaction of hormones and other factors in culture can lead to testable hypotheses of animal physiology, then resistance to culture studies in cellular endocrinology should vanish.

Tissue cultures have been a major tool in the elucidation of mammalian nucleic acid metabolism (Perry, 1962; Scherrer and Darnell, 1962; Weinberg and Penman, 1970). In these studies, investigators used HeLa cells that had been long established as heteroploid cultures and had been adapted to grow in suspension. The important principle that emerged is that biological macromolecules are synthesized as large precursors that are processed by cleavage to the final product. This principle was extended from RNA to proteins, resulting in the observation in cultured insulinomas that insulin was synthesized as a prohormone (Steiner *et al.*, 1967).

Cell cultures have also been used to good advantage in cytogenetics (Degrouchy, 1983; Edwards and McKusick 1982; Hsu, 1979; Tjio and Levin, 1956). Much that is known about chromosome localization of genes and chromosome anomalies in genetic diseases comes from mapping studies on cultured cells.

A direct application to humans was described by Goldstein, who spoke to the committee about his work on cultured human fibroblasts (Appendix E). Goldstein pointed out how these cells can be used in studying biological aging and age-dependent disease. If aging is basically a cellular process, then the study of cultured cells should provide information on the changes that occur as a cell changes in time. Genetic controls can also be most easily examined at a cell level.

In the workshop on diseases and aging (Appendix E), Weissman discussed the role of neutrophils in acute immunologic inflammatory response. Not only can neutrophil response be studied in cultured human cells, but it can also be modelled in dissociated cells of the sponge *Microciona prolifera*.

Several lines of evidence already support the notion that cell cultures are valid models for integrated physiology. Levi-Montalcini observed in the 1950s that tumor extracts stimulated nerve cell processes from sympathetic ganglia *in vitro* (for review see Levi-Montalcini, 1972). This observation was greeted with skepticism, since one could reasonably question the relevance of tumor extracts to the development of the sympathetic nervous system. During this period the importance of nucleic acids was becoming apparent. For this reason, Levi-Montalcini treated the tumor extracts with snake venom nucleases and observed that the venom preparation itself was active in promoting the outgrowth of nerve processes. She next reasoned that since the venom came from the salivary gland of a snake perhaps the salivary gland of the mouse would contain active material, and this turned out to be so. The prevailing attitude at that time was that tumor extracts, snake venom, salivary gland extracts, and tissue cultures could not be relevant to the normal development of the sympathetic nervous system. Antibodies to nerve growth factor were developed and injected into fetal animals. The animals were born without a sympathetic nervous system. Today no one doubts that this factor, which was first detected in culture experiments, is involved in the normal embryonic development of the sympathetic nervous system (Levi-Montalcini, 1972).

The platelet-derived growth factor (PDGF) was discovered in experiments in which serum, but not plasma, could support the growth of cultured cells (Balk *et al.*, 1973). It was traced by Ross to the platelets, and has come to assume great prominence in research on cancer and atherosclerosis (Ross *et al.*, 1974). PDGF is coded for by an oncogene, *V-sis*, and in turn activates a series of oncogenes in responsive cells (Doolittle *et al.*, 1983; Kelley *et al.*, in press). PDGF is also most probably involved in the generation of atherosclerotic lesions through its stimulation of the proliferation of smooth muscle cells in blood vessels to the extent that they occlude the vessel (Ross, 1983). The most convincing evidence for the physiological importance of PDGF comes from experiments in which antiserum to PDGF prevents the proliferation of smooth muscle where endothelial cells have been stripped (Ross *et al.*, 1974).

The increasing use of cell cultures as model systems is providing new approaches to many problems in biology.

THEORY (MATHEMATICAL MODELS)

It is difficult to isolate the role of theory in biological models because in the sense established by Popper (1959), all experiment requires tentative theory to make it legitimately science rather than random observation. Biology is thus motivated by a series of theories, but as these become more mathematical and more abstract, they begin to form a separate discipline, which requires specialized training to conduct its analytical procedures (see Burton, 1981; Eisenfeld and DeLisi, in press).

The workshop on cellular immunology did not include a theoretician; however, the theoretical work of Jerne in this field led to his being awarded a Nobel Prize in 1984 (Appendix A). In cellular immunology, several basic theories have been elaborated and specific questions have been formulated through use of mathematical models. For example, the logical options involved in analyzing idiotypical interactions in immune responses have been clarified. In cell-cell interactions, there is work going on to develop biophysical models of bridges between specific receptors. The perspectives on mathematical models in cellular immunology were further developed by Bell in the workshop on mathematical modelling (Appendix G). This presentation led to an understanding of some necessary relationships between theory and experiment.

The workshop on learning reflected a high level of mathematical analysis both in the specific papers and the general level of formal analysis necessary to approach this subject (Appendix C). Winston's discussion of artificial intelligence pointed to an area of mutual interest to theoreticians and neurobiologists interested in the mechanisms of learning, reasoning, and intelligence.

The use of macroscopic mathematical models in physiology and pharmacology has produced a rather large and well-developed accumulation of results. One example is the work discussed by Guyton, which is focused on models of the circulatory system with special reference to clinical hypertension. Here, according to Guyton, exciting new concepts are regularly discovered when running large-scale mathematical models on the computer (Appendix G).

At the opposite extreme, even simple models can be used to define more exactly parameters measuring metabolic states and to study glucose utilization in the elderly. This approach was outlined by Bergman (Appendix G) in his discussion of the minimal model approach and its application to aging and metabolic diseases. There are also formal models for pulmonary function, renal transport, and other functions. The use of pharmacokinetics to define clinical dosage regimens is well known, and more complete models yielding predictions of specific tissue levels are being developed for therapeutic delivery of novel drugs and also for developing better definitions of toxicological responses. This was discussed by Robert Dedrick in the workshop on mathematical modelling (Appendix G). In the workshop on models for the study of diseases and aging (Appendix E), the discussion of cardiac arrythmias by Glass illustrated mathematical modelling of an *in vitro* model system.

Development has long been subject to mathematical treatment (Nicolis and Prigogine, 1977; Thompson, 1917; Turing, 1952). An elegant current example of the mathematical modelling of morphogenesis was presented by O'Dell in a discussion of his work on the slime mold *Dictyostelium discoideum* (Appendix G).

Much mathematical modelling in biology is concerned with applications of physical chemistry to molecular structure, function, and aggregation. Charles DeLisi discussed computer-assisted studies of the relationships between molecular structure and biological activity (Appendix G), which exemplify the power of the computer to facilitate pattern recognition among vast numbers of objects. This well-established branch of biochemistry and biophysical chemistry has been fundamental to major advances in molecular biology.

In a sense, the process of assembling the matrix of biological information is a mathematical construct, and the search for connections in the matrix is, like most of science, an exercise in pattern recognition. Much work remains to be done on effective management of such large data bases. This is especially true for the collection of data through the use of modelling, as discussed in this report, because these studies generate a variety of data, e.g., observational and tabulated qualitative and quantitative data.

One common thread in the theory workshop was the necessary interplay between good mathematical modelling and experiment. Not only do experiments suggest mathematical models, but the models themselves often suggest further experimentation. Modelling without such interactive processes is unlikely to produce results that are as useful.

Most experimental biologists are most familiar with one type of mathematical modelling -- the *simulation* of data that have been collected. This has led to the somewhat negative view that mathematical models add little to understanding of the experimental data already in existence. However, formal models can be used in another manner -- a heuristic mode. In this usage, a model is a quantitative formulation of a hypothesis. In fact, a model that gives *incorrect* predictions is often useful, since it pinpoints what, in fact, is not as well known as presumed. The word quantitative is emphasized, since this is how much mathematical modelling is presently formulated; nonparametric methods under consideration may be useful for deriving qualitative conclusions. When used in this way, the model may provide a framework into which the experimental results must fit if a comprehensive picture is to emerge. Of course, simulations with a mathematical model are often useful in practice, such as in developing accurate definitions of the quantitative details of certain clinical practices such as drug administration.

Ways to improve the interaction between experimental and theoretical research were discussed at some length. One approach is to have one investigator learn both experimental techniques and mathematical modelling. The other is to ensure very close collaboration between two investigators with complementary strengths. Both arrangements were represented by workshop participants, and no agreement was reached concerning the best way to achieve the best interaction, except that more mathematics and modelling should be part of the regular biological curriculum and, conversely, more biology should be included in the regular curriculum for physical scientists and engineers interested in biological modelling.

To achieve the maximum benefits, mathematical modelling must be part of the experimental program from its inception. Otherwise, it may be difficult to analyze data because, for example, some condition was not held constant or some key variable was not measured. If done correctly, mathematical modelling can often increase the effectiveness of experiments by defining the key variables more precisely. This is especially true in complicated systems, since intuition is simply not adequate to the task. At times, even specific counterintuitive results can be suggested by the mathematical model. These models can sometimes aid in checking theories, resulting in more effective and economical experiments on biological specimens.

NIH SUPPORT FOR MODELS IN BIOMEDICAL RESEARCH

The NIH Division of Research Resources (DRR) provided the committee with unpublished data on the distribution of support for extramural research projects and the type of biological systems used in each. For the fiscal years 1977-1983, there was a slight decline from 27.5% to 22.9% in research dollars devoted to studies on humans, but a slight increase in the percentage of research dollars allocated to projects that involved the use of nonhuman mammals as research subjects. During the same period, the percentage of extramural research dollars that supported research on all other systems (e.g., invertebrates, nonmammalian vertebrates, cell and tissue culture systems, bacteria, viruses, and mathematical and computer simulations) remained essentially constant at about 29% of extramural research dollars, suggesting a slight shift in support away from human subjects and toward laboratory mammals (Table 4-1). When viewed as a percentage of total projects and subprojects, however, the distribution among human, laboratory mammal, and other subjects appears to be relatively constant at approximately 30%, 42%, and 26%, respectively.

A somewhat more detailed analysis was provided for fiscal years 1980-1983. Over this period the distribution of extramural research support remained essentially constant, in terms of percentage of total projects and subprojects (Table 4-2) as well as in terms of dollars (Table 4-3). Studies devoted solely to invertebrates received approximately 2% of the total support distributed to extramural grants; those devoted solely to nonmammalian vertebrates, 2.1%. Studies combining mammals and nonmammalian vertebrates, mammals and invertebrates, or

Table 4-1. Distribution of NIH Support of Extramural Research Among Humans, Laboratory Mammals, and Other Research Subjects, Expressed as Percentages of Total Dollars and of Total Projects and Subprojects[a]

Subject	Fiscal Year	Extramural Research Dollars, %	Total Projects and Subprojects, %
Humans	1977	27.5	32.4
	1978	26.8	31.2
	1979	26.8	29.2
	1980	25.0	28.9
	1981	23.8	29.7
	1982	23.2	31.5
	1983	22.9	32.2
Mammals	1977	43.5	41.9
	1978	44.0	42.5
	1979	44.9	43.8
	1980	45.0	44.2
	1981	47.3	44.1
	1982	48.1	43.5
	1983	47.9	42.7
Other[b]	1977	29.4	25.6
	1978	29.3	26.3
	1979	28.2	27.0
	1980	29.8	26.9
	1981	28.9	26.0
	1982	28.7	25.0
	1983	29.2	25.1

[a] Unpublished information provided by Division of Research Resources, National Institutes of Health.

[b] This category includes invertebrates, nonmammalian vertebrates, bacteria, viruses, mathematical and computer simulations, and other subjects.

Table 4-2. Distribution of NIH Support for Extramural Research, Expressed as Numbers of Projects Involving Various Categories of Research Subjects[a]

Research Classification[b]	1980 No. of Projects	1980 % of Projects	1981 No. of Projects	1981 % of Projects	1982 No. of Projects	1982 % of Projects	1983 No. of Projects	1983 % of Projects
H	10,033	30.3	10,116	30.6	10,534	32.2	11,458	32.9
HI	28	0.1	26	0.1	31	0.1	35	0.1
HM	3,695	11.2	3,528	10.7	3,449	10.5	3,572	10.3
HMI	56	0.2	70	0.2	98	0.3	84	0.2
HMN	219	0.7	209	0.6	188	0.6	163	0.5
HMNI	16	>0.01	12	>0.01	13	>0.01	8	>0.01
HN	65	0.2	60	0.2	54	0.2	64	0.2
HNI	1	>0.01	2	>0.01	12	>0.01	5	>0.01
I	739	2.2	725	2.2	672	2.1	698	2.0
N	9,541	28.8	9,535	28.9	9,279	28.4	9,748	28.0
MI	285	0.9	264	0.8	272	0.8	298	0.9
MN	666	2.0	662	2.0	661	2.0	701	2.0
MNI	59	0.2	58	0.2	85	0.3	58	0.2
N	786	2.4	758	2.3	774	2.4	783	2.3
NI	78	0.2	70	0.2	72	0.2	72	0.2
R	6,862	20.7	6,944	21.0	6,508	19.9	7,048	20.3
TOTAL	33,129		33,039		32,702		34,795	

[a] Unpublished information provided by Division of Research Resources, National Institutes of Health.

[b] H = Humans; I = Invertebrates; M = Nonhuman mammals; N = Nonmammalian vertebrates; R = Other research systems (e.g., bacteria, viruses, plants, biological preparations, cell and organ culture, mathematical systems).

Table 4-3. Distribution of NIH Support for Extramural Research, Expressed as Dollar Amounts Allocated to Various Categories of Research Subjects[a]

Research Classification[b]	1980 Dollars Awarded	1980 % of Total Dollars	1981 Dollars Awarded	1981 % of Total Dollars	1982 Dollars Awarded	1982 % of Total Dollars	1983 Dollars Awarded	1983 % of Total Dollars
H	766,758,349	27.7	719,713,783	25.3	693,097,446	24.3	783,661,178	24.2
HI	2,785,845	0.1	2,445,786	0.1	3,708,152	0.1	3,895,940	0.1
HM	332,299,860	12.0	345,863,175	12.2	351,177,036	12.3	400,158,256	12.4
HMI	5,528,593	0.2	7,629,089	0.3	9,807,186	0.3	9,997,878	0.3
HMN	19,770,276	0.7	20,601,039	0.7	20,485,601	0.7	19,571,561	0.6
HMNI	1,823,613	0.1	1,643,156	0.1	1,715,764	0.1	1,336,877	>0.01
HN	4,998,800	0.2	5,236,600	0.2	5,526,560	0.2	6,960,610	0.2
HNI	99,631	>0.01	173,393	>0.01	1,227,086	>0.01	>755,201	0.01
I	53,379,256	1.9	56,370,010	2.0	54,564,121	1.9	64,093,202	2.0
M	765,535,348	27.7	825,844,280	29.0	849,065,134	29.8	962,667,804	29.8
MI	20,305,722	0.7	21,669,277	0.8	25,278,657	0.9	29,795,939	0.9
MN	57,058,730	2.1	61,496,736	2.2	66,901,880	2.3	74,260,448	2.3
MNI	6,028,938	0.2	6,294,183	0.2	8,608,568	0.3	7,301,142	0.2
N	54,352,594	2.0	58,816,887	2.1	60,977,113	2.1	69,123,495	2.1
NI	6,403,445	0.2	7,494,069	0.3	7,703,005	0.3	8,180,269	0.3
R	671,294,725	24.2	702,314,379	24.7	692,789,442	24.3	792,977,894	24.5
TOTAL	2,768,423,725		2,843,605,842		2,852,632,751		3,234,737,694	

[a] Unpublished information provided by Division of Research Resources, National Institutes of Health.

[b] H = Humans; I = Invertebrates; M = Nonhuman mammals; N = Nonmammalian vertebrates; R = Other research systems (e.g., bacteria, viruses, plants, biological preparations, cell and organ culture, mathematical systems).

nonmammalian vertebrates and invertebrates together received approximately 4.8%. Support for research on nonanimal models (e.g., microorganisms or mathematical models) accounted for 24.2% in FY 1980, 24.7% in FY 1981, 24.3% in FY 1982, and 24.5% in FY 1983, respectively, while total support for projects involving nonhuman mammals rose from 45% in FY 1980 to an average of 47.7% for FY 1981 through FY 1983.

In drawing inferences from these data it is necessary to exercise caution. The cost-effectiveness of research on various taxa are difficult to compare meaningfully, since it is highly unlikely that any two research projects are pursuing exactly the same objective. It can be readily seen from these figures that the proportion of resources allocated to nonmammalian animals is quite small. The committee could not determine the reasons for this, but as noted below, the contribution of nonmammalian vertebrates and invertebrates to biomedical research seems to have been much greater than the support they now enjoy in NIH-sponsored extramural programs.

Additional data provided by NIH for FY 1983 included the distribution of extramural support across its bureaus, institutes, and divisions (BIDs), the model systems used, and the problems investigated. These data indicate that the use of nonmammalian models and model systems is widespread among the BIDs. Twelve BIDs supported a total of 182 projects devoted to human cells or tissues in culture with a value of $16,594,111. Twelve BIDs supported research projects in which cells or tissues from sources other than humans were used as models in 335 projects at a cost of $30,778,442. The distribution of support for research devoted exclusively to invertebrates totaled $125,338,218 across 1,341 projects or subprojects. Of these, 320 at $28,079,763 involved invertebrates as pathogens.

Eight BIDs supported a total of 23 projects in which mathematical models were used to study a variety of biomedical research problems, including the topics covered in the workshop on mathematical modelling (Appendix F). During FY 1983, support for mathematical modelling was $1,172,000, and computer simulations were being used in 54 projects funded at $5,981,000.

The distribution of support among various taxa of invertebrates is shown in Table 4-4. NIH resources are invested in many different applications of a variety of invertebrate systems across a broad spectrum of problems in biology and medicine. Compared to the numbers and variety of invertebrate species and the potential for newly studied species to provide valuable new model systems, however, the support for invertebrate systems appears modest. As noted earlier in this chapter, the potential for research on invertebrates appears to be considerable, based on the numbers of species and on the rich payoff from research conducted on invertebrates to date. An examination of organisms used in Nobel Prize-winning research supports this view (Appendix A). Because the compilation of these data was somewhat arbitrary and the data are

Table 4-4. Distribution of NIH Support for Research Using Invertebrates, for Fiscal Year 1983.

System	Number of Projects Citing Use[a]	Number of BIDs[b] Supporting Use
Protozoa	202	11
Porifera	1	1
Coelenterata	8	5
Platyhelmintha[c]	55	2
Nemata (nematodes)	59	7
Mollusca		10
Aplysia sp.	35	
Cephalopoda	36	
Gastropoda	24	
Other	46	
Total	141	
Annelida	27	7
Arthropoda		10
Crustacea	55	
Drosophila sp.	235	
Other insecta	178	
Limulus sp.	18	
Other arthropoda	29	
Total	515	
Echinodermata		5
Echinoidea (sea urchins)	46	
Other	41	
Total	87	
Miscellaneous invertebrates[d]	246	11
TOTAL	1,341	

[a]Unpublished data provided by Division of Research Resources, National Institutes of Health.

[b]BID = Bureaus, institutes, and divisions of the National Institutes of Health.

[c]Projects devoted largely to research on parasitic diseases.

[d]Includes projects cited as "invertebrate" or "invertebrate animal."

not necessarily precise, the information should be regarded as no more than suggestive.

Although much more detailed analyses based on accounts from the Nobel Prize winners themselves would be required to determine the precise reasons for their choice of research materials and strategies for interpreting data, one general observation can be made. The Nobel laureates appear to have made greater use of nonmammalian vertebrates, invertebrates, and combinations of mammals and nonmammalian vertebrates than have investigators supported by the NIH in recent years.

Nobel Prizes are awarded retrospectively, whereas NIH grant awards are presumably prospective. Thus, whereas Nobel Prize committees benefit from a nomination process and a retrospective evaluation of the nominee's entire research record, the NIH must anticipate the potential yield of a proposed program of research before awarding grants -- presumably a much more difficult task. Nevertheless, the pattern of Nobel awards may provide some general guidance to the NIH awards process, which may be undervaluing model systems likely to provide significant opportunities for both one-to-one and many-to-many modelling.

It should be possible for the NIH administration to determine whether investigators using nonmammalian systems have more difficulty getting support than investigators using mammalian systems and to correct any imbalance that might be discovered. But even if there is no imbalance, it is nevertheless possible that the present balance does not reflect the preference of all investigators. Of necessity, investigators select their model systems for investigation in part according to the perceived availability of support. Therefore, NIH might consider encouraging interest in the development of nonmammalian systems through postdoctoral fellowships, symposia, and direct support of model development. Development of nonmammalian systems might also be encouraged by providing to investigators in computer-accessible form elements of the matrix of biological knowledge as described in the following chapter.

HIGH CONNECTIVITY MODELS

As the body of knowledge concerning an organism or biological preparation increases in breadth, depth, and variety, each new increment of data can be evaluated against a larger background, or transferred to a larger matrix of information. Furthermore, new information about one system can be related not only to other data on that system but also to other systems. The matrix of biological information onto which new information is being transferred includes other organisms, of the same or different taxa and of the same or a different level of organization. The opportunities to exploit a set of data are proportional to the size, quality, and organization of the matrix of preexisting information to which the new data set will be transferred (or onto which it will be mapped). The ability to map the information derived from an experimental system increases as the body of knowledge about the particular organism itself increases. The more that is learned about a system, the more abundant are the opportunities for many-to-many modelling from that system to other taxa.

Where the corpus of knowledge about a system is large, that is, when the investigator starts with a substantial understanding of the functions of that system, the opportunities for many-to-many modelling are maximized. Consequently, the opportunities for new insights into biological mechanisms are maximized. Such organisms may be regarded as high connectivity models, i.e., there is a great potential that connections will be made between observations on the model system and data on other systems. In other words, there is a potential for making generalizations from information generated with the model system, according to Margenau (1977; see Chapter 3).

Examples of high connectivity models include many widely studied and well-understood invertebrates and microorganisms, such as the bacterium *Escherichia coli* and the yeast *Saccharomyces cerevisiae*, the slime mold *Dictyostelium*, the algae *Chlorella* and *Chlamydomonas*, the protozoan *Paramecium*, the fungus *Neurospora*, the marine dinoflagellate *Gonyaulax*, the nematode *Caenorhabditis elegans*, and the fruit fly *Drosophila*.

The extensive knowledge about these organisms permits us to understand better their life cycle, genetics, nutrition, culture requirements, energy metabolism, and other biological features. In addition, the organisms are generally well understood taxonomically and phylogenetically. For some of them, genetic fine structure is being elucidated. The connectivity of many systems would be improved and their value for modelling increased if efforts were made to close conspicuous gaps in the data base about the model itself, especially when many aspects of a system are well understood.

The nematode *C. elegans* is an example of a system with significant potential for high connectivity. As described in Appendix F, the genetics of this organism is well understood, many mutants are available, and the entire neuroanatomy is known. In addition, the lineage of every cell of *C. elegans* can now be traced back to the zygote. Thus for the first time we know the origin of every cell in a multicellular organism. In and of itself, this knowledge of cell lineage provides interesting and presumably important information about development; some of the lineages are unexpected, as are many of the migrations that cells undergo during development of the mature organism. As more is learned about other aspects of the biology of *C. elegans*, there will be more opportunities to relate control of cell lineage to genetics, metabolism, behavior, fine structure, and all the other functions and processes that constitute the organism's total biology. *C. elegans* is rapidly gaining attention as a potential model for various biological functions.

Sometimes specific features of an organism make it a useful model independent of or in the absence of knowledge of the rest of its biology. For example, the study of fertilization in sea urchins has been profitable without knowledge of sea urchin physiology or genetics.

Of course, there are limits to the applicability of organisms such as these as models for functions in complex multicellular organisms. Their very simplicity limits their value with respect to higher nervous function, for example. Mammalian models, including nonhuman primates and humans themselves, remain uniquely important for a whole range of problems.

TOXICOLOGY AND TOXICITY TESTING

Perhaps nowhere in biomedical research has the search for new and improved model systems been pursued with greater vigor than in toxicology and toxicity testing (Dangai, 1983). The ever-growing need to assess the risk of chemicals has been accompanied by a desire to increase sensitivity and accuracy of testing procedures, to reduce the time and money required to conduct tests, and to reduce the numbers of mammals required. As a consequence, the development of new and improved systems for investigating the toxic effects of chemicals is pursued with increased support; it has become an end in itself and is given high priority in both the public and private sectors. The high cost of testing (full life cycle rodent tests for carcinogenicity may cost more than a half million dollars per chemical) coupled with the federal laws regulating chemical exposures has spurred the search for faster, cheaper, more accurate alternatives to the use of mammals in tests to explore the potential adverse effects of environmental chemicals.

Toxicology and toxicity testing are areas in which cell and tissue culture techniques have been vigorously exploited, although extrapolation to the intact system or to another organism may be especially difficult. In toxicology, as in other disciplines, cell and tissue culture systems can contribute important information about components of more complex systems (Nardone, 1979; Wilson, 1978).

In recent years, a substantial amount of new data has been produced on systems for toxicity testing. The proceedings of major conferences and symposia on this topic have been published both in England (Balls *et al.*, 1983) and in the United States (National Institutes of Health, 1981).

Because of the attention that alternative systems for toxicity testing have been receiving elsewhere, the committee chose not to emphasize them in this study; however, it does not wish to diminish the relevance of toxicology and toxicity testing models to models for biomedical research in general. On the contrary, it recognizes that normal biological processes are often elucidated by the study of abnormalities, including toxic effects. Thus, studies of the inhibitory effects of arsenic, fluoride, cyanide, and heavy metals have helped to reveal the enzymatic pathways of carbohydrate metabolism and cellular respiration, and studies of colchicine, a mitosis inhibitor, have helped to explain the assembly and disassembly of subunits of the cytoplasmic matrix.

Model development in toxicology can therefore be expected to result in data and systems of value in other areas of biomedical research, and the general research community will benefit from improved bioassays for toxic effects in humans.

Researchers in many areas seemingly unrelated to toxicology might contribute significantly to the search for new and improved models for toxicity by examining the effects of exogenous chemicals on the systems they investigate. Through an appropriate clearinghouse, it should be possible to place biologists working with diverse organisms in touch with toxicologists who are trying to characterize the biological effects of environmental chemicals and arrange to supply the chemical to the model user along with relevant toxicity data. A mechanism would be needed for matching model users with particular chemicals. The NIH might explore the feasibility of designing a procedure to accomplish this task, possibly facilitated by computerized data bases. By testing a chemical of interest to environmental toxicologists, investigators using novel systems to study, for example, development, neurophysiology, gene expression, diseases, aging, learning, regulation, or cellular immunology may aid in the characterization of the health effects of the chemicals and may even contribute new bioassays to assess the potential health hazards of chemicals in general.

As a hypothetical but by no means unrealistic example, one might imagine the outcome if workshop participants were asked to examine the effects of a particular chemical on the biological function that they are studying in the model systems they are currently using. If the same chemicals were examined by all participants, the resulting data would characterize the effects of that chemical on a number of biological activities in a variety of systems and should help to elucidate its mechanisms of toxicity.

Very few substances have been tested in more than one species, and there has been insufficient rationale for selecting the species (usually the mouse, rat, rabbit, guinea pig, or monkey) on which tests will be performed. Only by building up a body of theory through the collection of data on a variety of systems will it be possible to project in what way the human is likely to react to the chemical (Lederberg, 1981a,b). The clearinghouse function, through which new combinations of chemicals and systems would be tested, would constitute an important stimulus to the development of comparative toxicology as a discipline close, if not central, to general comparative biology.

Theory building based on comparative toxicology is not entirely neglected. The search for principles of extrapolation from model system to human is based on both one-to-one and many-to-many modelling (Calabrese, 1983).

SUMMARY

Whereas the use of nonmammalian vertebrates, invertebrates, and microorganisms as model systems derives from homology based on shared evolution and phylogenetic relationships, the other types of models considered by the committee, i.e., mathematical and some cell and tissue culture models, are strategies based directly on analogy. As strategies they contain an inherent test of worth. They either work (i.e., predict outcomes), or they are modified or abandoned. They are often heuristic. Used properly with good methodology, they justify themselves and require no further defense.

The committee has devoted much of its attention and discussion to invertebrates and nonmammalian vertebrates, because it saw among the numerous invertebrate and nonmammalian vertebrate species a significant potential for the discovery of additional important homologs and analogs to biological processes in general and to human structure and function, both normal and diseased, in particular. Although the committee would hardly argue for distribution of research support as a function of numbers of species in different major taxa, the vast number of invertebrate and nonmammalian vertebrate taxa, compared to the number of mammalian taxa, suggests that the former deserve more attention. There is, for example, enormous diversity among fish: a number of classes, scores of orders, and more species than all the mammals, birds, reptiles, and amphibians together. Further study of representatives of major taxa will shed light on important relationships and provide insights that will greatly increase the interconnected components of the matrix of biological knowledge.

In making the argument that more attention and support should be given to the search for biomedical models among invertebrates and nonmammalian vertebrates, the committee does not intend to dismiss the possibility that new models may be derived from microorganisms. On the contrary, the power of microbial genetics as a model for the basic molecular genetic machinery of all living things has been so strongly established and widely exploited that it should make further justification unnecessary. Nor does the committee intend to slight the plant kingdom, which also should require no further justification as a source of models not only for uniquely plant functions such as photosynthesis but also for universal cell functions.

Although not specifically asked to do so, the committee has considered nonmammalian vertebrates in some cases and mammals, including primates, in others. They were included in this study because the advantages of nonmammalian systems must be judged in comparison to mammals. For many problems, the best model is the species closest to the human in terms of its taxonomy and physiology. This is especially true for higher, integrative levels of function, such as aspects of learning, and for some clinical applications, such as surgery. Although it is often possible to subdivide a system and to find models for its components in the best tradition of reductionism, it is also often necessary to verify one's intuitions regarding the appropriateness of a model with observations on the best available close homolog -- a primate. Also, nonmammalian models are not always useful or adequate. When there are no alternative models or when data and conclusions derived from simpler models need verification, the use of mammals must continue.

These conclusions, and indeed the entire rationale for this report, are based on a perception that emerged during the course of the workshop series. That is, with the exception of some aspects of the modelling of disease states, modelling in biomedical research is many-to-many, i.e., to the matrix of biomedical knowledge, rather than one-to-one, i.e., directly to the human. The matrix of biological knowledge is interrelated because of shared evolution and the consequent opportunities to exploit homologs in the search for good analogs. This emerging new perception is described in somewhat greater detail in the next chapter.

REFERENCES

Balk, S. D., J. F. Whitfield, T. Youdale, and A. D. Braun. 1973. Roles of calcium, serum, plasma, and folic acid in the control of proliferation of normal and rous sarcoma virus-infected chicken fibroblasts. Proc. Natl. Acad. Sci. USA 70:675-679.

Balls, M., R. J. Riddell, and A. N. Worden, eds. 1983. Animals and Alternatives in Toxicity Testing. Academic Press, London. 550 pp.

Bargiello, T. A., and M. W. Young. 1984. Molecular genetics of a biological clock in *Drosophila*. Proc. Natl. Acad. Sci. USA 81:2142-2146.

Barnes, D., and G. Sato. 1980. Serum free cell culture: A unifying approach. Cell 22:649-655.

Bender, W., M. Akam, F. Karch, P. A. Beachy, M. Peifer, P. Spierer, E. B. Lewis, and D. S. Hogness. 1983. Molecular genetics of the biothorax complex in *Drosophila melanogaster*. Science 221:23-29.

Bennett, M. V. L. 1977. Electrical transmission: A functional analysis and comparison to chemical transmission. Pp. 35-416 in S. R. Geiger, ed. Handbook of Physiology. The Nervous System, Sec. 1., Cellular Biology of Neurons, Part 1, Vol. I. Williams and Williams, Baltimore, Maryland.

Bowen, M. F., D. S. Saunders, W. E. Bollenbacher, and L. I. Gilbert. 1984. *In vitro* reprogramming of the photoperiodic clock in an insect brain-retrocerebral complex. Proc. Natl. Acad. Sci. USA 81:5881-5884.

Brecha, N. C., W. Eldred, R. O. Kuljis, and H. J. Karten. 1984. Identification and localization of biologically active peptides in the vertebrate retina. Prog. Ret. Res. 3:185-226.

Burton, T. A., ed. 1981. Mathematical Biology: A Conference on Theoretical Aspects of Molecular Science. Pergamon Press, New York. 249 pp.

Calabrese, E. J. 1983. Principles of Animal Extrapolation. Wiley-Interscience, New York. 603 pp.

Cohen, S. In press. Murakami, ed. International Symposium on Growth and Differentiation of Cells in Defined Environment. Fukuoka, Japan. Springer-Verlag, New York.

Dangai, R. 1983. Alternative methods could cut animal use in toxicity tests. Chem. Eng. News. 61:7-13.

Degrouchy, J. 1983. The human gene map: A review. Biomed. Pharmacother. 37:159-172.

Doolittle, R. F., M. W. Hunkapiller, L. E. Hood, S. G. Devare, K. C. Robbins, S. A. Aaronson, and H. N. Antoniades. 1983. Simian sarcoma virus _onc_ gene, V-_sis_, is derived from the gene (or genes) encoding a platelet-derived growth factor. Science 221:275-277.

Dowling, J. E. 1979. Information processing by local circuits: The vertebrate retina as a model system. Pp. 163-181 in F. O. Schmitt and F. G. Worden, eds. The Neurosciences, Fourth Study Program. MIT Press, Cambridge, Massachusetts.

Edwards, J. H., and V. A. McKusick. 1982. Report of the Committee on Unassigned Linkage Groups. Cytogenet. Cell Genet. 32:191-193.

Eisenfeld, J., and C. DeLisi. In press. Mathematics and Computers in Biomedicine. Elsevier North-Holland, Amsterdam, the Netherlands.

Ephrussi, B. 1942. Chemistry of "eye color hormones" of _Drosophila_. Q. Rev. Biol. 17:327-338.

Epple, A. 1969. The endocrine pancreas. Pp. 275-319 in W. S. Hoar and D. J. Randall. eds. Fish Physiology, Vol. 2. Academic Press, New York.

Fisher, H. W., T. T. Puck, and G. Sato. 1958. Molecular growth requirement of single mammalian cells: The action of fetuin in promoting cell attachment to glass. Proc. Natl. Acad. Sci. USA 44:4-10.

Fraenkel, G. 1948. B_T, a new vitamin of the B-group and its relation to the folic acid group and other anti-anemia factors. Nature 161:981-983.

Frankenhaeuser, B., and A. F. Huxley. 1964. The action potential in myelinated nerve fibre of _Xenopus laevis_ as computed on the basis of voltage clamp data. J. Physiol. 171:302-315.

Fujita, S. C., S. L. Zipursky, S. Benzer, A. Ferrus, and S. L. Shotwell. 1982. Monoclonal antibodies against the _Drosophila_ nervous system. Proc. Natl. Acad. Sci. USA 79:7929-7933.

Gelperin, A., and N. Culligan. 1984. _In vitro_ expression of _in vivo_ learning by an isolated molluscan CNS. Brain Res. 304:207-213.

Goodman, C. S., M. J. Bastiani, C. Q. Doe, S. duLac, S. L. Helfand, J. Y. Kuwada, and J. B. Thomas. 1984. Cell recognition during neuronal development. Science 225:1271-1279.

Gordon, A. M., A. F. Huxley, and F. J. Julian. 1966. The variation in isometric tension with sarcomere length in vertebrate muscle fibres. J. Physiol. 184:170-192.

Gurdon, J. B., and R. A. Laskey. 1970. The transplantation of nuclei from single cultured cells into enucleate frogs' eggs. J. Embryol. Exp. Morphol. 24:227-248.

Harrison, R. G. 1907. Observations on the living developing nerve fiber. Proc. Soc. Exp. Biol. Med. 4:140-143.

Hayashi, I., and G. Sato. 1976. Replacement of serum by hormones permits growth of cells in a defined medium. Nature 259:132-134.

Hodgkin, A. L., and A. F. Huxley. 1952. A quantitative description of membrane current and its application to conduction and excitation in nerve. J. Physiol. 117:500-544.

Hoffmann, J., and M. Lagueux. In press. Endocrinological aspects of embryogenesis. Pp. 435-460 in G. Kerkut and L. I. Gilbert, eds. Comprehensive Insect Physiology, Biochemistry and Pharmacology. Endocrinology I, Vol. I. Pergamon Press, New York.

Holley, R. W., and J. A. Kiernan. 1974. Control of the initiation of DNA synthesis in 3T3 cells: Serum factors. Proc. Natl. Acad. Sci. USA 71:2908-2911.

Hopkins, F. G. 1895. The pigments of the pteridae: A contribution to the study of excretory substances which function in ornament. Philos. Trans. R. Soc. London, Ser. B. 186:661-682.

Hotta, Y., and S. Benzer. 1972. Mapping of behavior in _Drosophila_ mosaics. Nature 240:527-535.

Hsu, T. C. 1979. Human and Mammalian Cytogenetics: An Historical Perspective. Springer-Verlag, New York. 186 pp.

Huxley, H., and J. Hanson. 1954. Changes in the cross-striations of muscle during contraction and stretch and their structural interpretation. Nature 173:973-976.

Huxley, H. E., R. M. Simmons, A. R. Faruqi, M. Kress, J. Bordas, and M. H. J. Koch. 1981. Millisecond time-resolved changes in x-ray reflections from contracting muscle during rapid mechanical transients, recorded using synchrotron radiation. Proc. Natl. Acad. Sci. USA 78:2297-2301.

Jacquez, J. A., and E. Barry. 1951. Tissue culture media: The essential non-dialyzable factors in human placental cord serum. J. Gen. Physiol. 34:765-774.

Jeffery, W. R., C. R. Tomlinson, and R. D. Brodeur. 1984. mRNA localization in the myoplasm of ascidian embryos. Pp. 145-163 in E. H. Davidson and R. A. Firtel, eds. Molecular Biology of Development. UCLA Symposia on Molecular and Cellular Biology, New Series, Vol. 19. Alan R. Liss, New York.

Kandel, E. R., T. Abrams, L. Bernier, T. J. Carew, R. D. Hawkins, and J. H. Schwartz. 1983. Classical conditioning and sensitization share aspects of the same molecular cascade in *Aplysia*. Cold Spring Harbor Symp. Quant. Biol. 48:821-830.

Katz, G. 1969. The Release of Neural Transmitter Substances. Sherrington Lectures, No. 10. Charles C Thomas, Springfield, Illinois. 60 pp.

Keilin, D. 1925. On cytochrome, a respiratory pigment common to animals, yeast and higher plants. Proc. R. Soc. London, Ser. B. 98:312-339.

Kelley, K., P. Leder, and C. D. Stiles. In press. Murakami, ed. International Symposium on Growth and Differentiation of Cells in Defined Environment. Fukuoka, Japan. Springer-Verlag, New York.

Kramer, K. In press. Vertebrate hormones in insects. Pp. 511-536 in G. Kerkut and L. I. Gilbert, eds. Comprehensive Insect Physiology, Biochemistry and Pharmacology. Embryogenesis and Reproduction, Vol. 7. Pergamon Press, New York.

Lane, C. D. 1981. The fate of foreign proteins introduced into *Xenopus* oocytes. Cell 24:281-282.

Lederberg, J. 1981a. A challenge for toxicologists. Regul. Toxicol. Pharmacol. 1:110-112.

Lederberg, J. 1981b. Comparative toxicology, environmental health and national productivity. Am. J. Med. 70:9-11.

Le Roith, D., J. Shiloach, J. Roth, and M. A. Lesniak. 1980. Evolutionary origins of vertebrate hormones: Substances similar to mammalian insulins are native to unicellular eukaryotes. Proc. Natl. Acad. Sci. USA 77:6184-6188.

Le Roith, D., J. Shiloach, M. Berelowitz, L. A. Frohman, A. S. Liotta, D. T. Krieger, and J. Roth. 1983. Are messenger molecules in microbes the ancestors of the vertebrate hormones and tissue factor. Fed. Proc. 42:2602-2607.

Lettvin, J. Y., H. R. Maturana, W. S. McCulloch, and W. H. Pitts. 1959. What the frog's eye tells the frog's brain. Proc. Inst. Radio Eng. 47:1940-1951.

Levi-Montalcini, R. 1972. The nerve growth factor. Pp. 25-45 in G. Steiner and E. Schonbaum, eds. Immunosympathectomy. Elsevier, New York.

Levintow, L., and H. Eagle. 1961. Biochemistry of cultured mammalian cells. Ann. Rev. Biochem. 30:605-640.

Lipmann, F. 1971. Wanderings of a Biochemist. Wiley Interscience, New York. 229 pp.

Loewi, O. 1933. The humoral transmission of nervous impulse. Harvey Lectures, Ser. 28:218-233.

Loose, D. S., E. P. Stover, A. Restrepo, D. A. Stevens, and D. Feldman. 1983. Estradiol binds to a receptor-like cytosol binding protein and initiates a biological response in *Paracoccidioides brasiliensis*. Proc. Natl. Acad. Sci. USA 80:7659-7663.

Macleod, J. J. R. 1922. The source of insulin. J. Metab. Res. 2:149-172.

Margenau, H. 1977. The Nature of Physical Reality: A Philosophy of Modern Physics. Ox Bow Press, Woodbridge, Conn. 467 pp.

McGinnis, W., M. S. Levine, E. Hafen, A. Kuroiwa, and W. J. Gehring. 1984. A conserved DNA sequence in homeotic genes of the *Drosophila* antennapedia and bithorax complexes. Nature 309:428-933.

Miller, J. A. 1984. Controlling the body geometric. BioScience 34:548.

Miller, C. A., and S. Benzer. 1983. Monoclonal antibody cross-reactions between *Drosophila* and human brain. Proc. Natl. Acad. Sci. USA 80:7641-7645.

Nardone, E. 1979. Tissue culture systems in toxicity testing. Pp. 91-101 in A. N. Rowan and C. J. Stratmann, eds. The Use of Alternatives in Drug Research. Macmillan, London, England.

National Institutes of Health. 1981. Proceedings of the Symposium on Trends in Bioassay Methodology: *In Vivo*, *In Vitro*, and Mathematical Approaches. NIH Publication No. 82-2382. National Institutes of Health, Bethesda, Maryland. 371 pp.

Nicolis, G., and I. Prigogine. 1977. Self-organization in non-equilibrium systems. Wiley-Interscience, New York. 491 pp.

Nishizuka, Y. 1983. Phospholipid degradation and signal translation for protein phosphorylation. Trends Biochem. Sci. 8:13-16.

Perry, R. P. 1962. The cellular sites of synthesis of ribosomal and 4S RNA. Proc. Natl. Acad. Sci. USA 48:2179-2186.

Piccolino, M., J. Neyton, and H. M. Gerschenfeld. 1984. Decrease of gap junction permeability induced by dopamine and cyclic adenosine 3':5'-monophosphate in horizontal cells of turtle retina. J. Neurosci. 4:2477-2488.

Polyak, S. 1957. The Vertebrate Visual System. University of Chicago Press, Chicago, Illinois. 1390 pp.

Popper, K. R. 1959. The Logic of Scientific Discovery. Basic Books, New York. 480 pp.

Prichard, R. W. 1976. Animal models in human medicine. Pp. 169-172 in Animal Models of Thrombosis and Hemorrhagic Disease. Publication No. (NIH) 76-982. U. S. Department of Health, Education, and Welfare, Washington, D.C.

Raff, R. A., H. V. Colot, S. E. Selvig, and P. R. Groes. 1972. Oogenetic origin of messenger RNA for embryonic synthesis of microtubule proteins. Nature 235:211-214.

Ross, R. 1983. Atherosclerosis: A question of endothelial integrity and growth control of smooth muscle. Harvey Lectures 77:161-182.

Ross, R., J. Glomset, B. Kariya, and L. Harker. 1974. A platelet-dependent serum factor that stimulates the proliferation of arterial smooth muscle cells *in vitro.* Proc. Natl. Acad. Sci. USA 71:1207-1210.

Roth, T. F., and K. R. Porter. 1964. Yolk protein uptake in the oocyte of the mosquito *Aedes aegypti* (L). J. Cell Biol. 20:313-332.

Rubin, G. M., and A. C. Spradling. 1982. Genetic transformation of *Drosophila* with transposable element vectors. Science 218:348-353.

Sato, G. 1975. The role of serum in cell culture. Pp. 391-396 in G. Witlack, ed. Biochemical Action of Hormones, Vol. III. Academic Press, New York.

Sato, G., G. Augusti-Tocco, M. Posner, and P. Kelly. 1970. Hormone-secreting and hormone-responsive cell cultures. Recent Progr. Horm. Res. 26:539-546.

Schaller, H. C., and H. Bodenmuller. 1984. Structure and function of the head activator in *Hydra* and in mammals. Paper presented at a Symposium on Comparative Aspects of Opioid and Related Mechanisms. State University of New York, May 1984.

Scheller, R. H., R. R. Kaldany, T. Kreiner, A. C. Mahon, J. R. Nambu, M. Schaefer, and R. Taussig. 1984. Neuropeptides: Mediators of behavior in *Aplysia*. Science 225:1300-1308.

Scherrer, K., and J. E. Darnell. 1962. Sedimentation characteristics of rapidly labelled RNA from HELA cells. Biochem. Biophys. Res. Comm. 7:486-490.

Spradling, A. C., and G. M. Rubin. 1982. Transposition of cloned P elements into *Drosophila* germ line chromosomes. Science 218:341-347.

Steiner, D. F., D. Cunningham, L. Spigelman, and B. Aten. 1967. Insulin biosynthesis: Evidence for a precursor. Science 157:697-700.

Suchanek, G., G. Kreil, and M. A. Hermodson. 1978. Amino acid sequence of honeybee prepromelittin synthesized *in vitro*. Proc. Natl. Acad. Sci. USA 75:701-704.

Sulston, J. E., E. Schierenberg, J. G. White, and J. N. Thomson. 1983. The embryonic cell lineage of the nematode *Caenorhabditis elegans*. Dev. Biol. 100:64-119.

Tata, J. R. 1984. The action of growth and developmental hormones-evolutionary aspects. Pp. 1-58 in R. F. Goldberger and K. R. Yamamoto, eds. Biological Regulation and Development. Hormone Action, Vol. 3B. Plenum Press, New York.

Temin, H., R. W. Pierson, Jr., and N. C. Dulak. 1972. The role of serum in the control of multiplication of avian and mammalian cells in culture. Pp. 49-81 in G. H. Rothblat and V. J. Cristofalo, eds. Growth, Nutrition and Metabolism of Cells in Culture, Vol. I. Academic Press, New York.

Thompson, D. 1917. On Growth and Form. Cambridge University Press, Cambridge, England. 793 pp.

Timberlake, W. E., and W. C. Orr. 1984. Steroid hormone regulation of sexual reproduction in *Achlya*. Pp. 255-283 in R. F. Goldberger and K. R. Yamamoto, eds. Biological Regulation and Development. Hormone Action, Vol. 3B. Plenum Press, New York.

Tjio, J. H., and A. Levin. 1956. The chromosome number of man. Hereditas 42:1-6.

Turing, A. M. 1952. The chemical basis of morphogenesis. Phil. Trans. R. Soc. London, Ser. B. 237:37-72.

Weinberg, R. A., and S. Penman. 1970. Processing of 45s nucleolar RNA. J. Mol. Biol. 47:169-178.

Wilson, J. G. 1978. Review of *in vitro* systems with potential for use in teratogenicity screening. J. Exp. Pathol. Toxicol. 2:149-167.

5

A New Perspective on the Role of Models

The broad applicability of modelling relationships perceived by the committee provides a new perspective in the search for models in biomedical research. Since the explicit statement of this relationship is not familiar to most investigators, some background is provided in the following paragraphs.

The theoretical structure of classical physics consists of a few postulates, such as Newton's law, the laws of thermodynamics, and Maxwell's laws. From these basic constructs explanations of all experimental phenomena within the domains encompassed by the theoretical structure may be developed. In this regard, physics has been held out as the model for all of science.

Biology, however, is different. It lacks such universal postulates, but it does possess a number of generalizations whose validity seems to rest on evolutionary relationships. Once a process or structure with survival value emerges, it is used over and over again across a broad phylogenetic domain. The differences between the conceptual framework of biology and physics have been discussed in some detail by Schaffner (1980).

The idea of biological generalizations first set forth by Aristotle was formalized in cell theory in the mid-1830s. Since that time there has been a steady increase in knowledge of organisms, including new information on commonalities among structural features, among biochemical elements, and among strategies for different processes. Following is a discussion of some of the generalizations discovered.

One set of atomic components organized into approximately 1,000 low molecular weight molecules make up the overwhelming mass of all living organisms. The sequences in which these molecules are biochemically transformed, i.e., their metabolic pathways, are similar among all organisms. That is, the intermediary metabolism of any of the several millions of species of organisms is a subset of one master chart, e.g., the Chart of Intermediary Metabolism (Sallach, 1972). Of course, there are major differences among taxa in their dependence on the tricarboxylic acid cycle, on aerobic pathways, and even on oxygen-based energy. Nonetheless, the chart is a very general model of all metabolism and serves as a prelude to the kind of relationship described below. Also common to all organisms is the method of macromolecule synthesis and the storage of genetic information, as well as the genetic code and the cellular hardware used in polymer formation. Within each major taxon, not only cell types but also most cell organelles are ubiquitous. Thus the modular units out of which organisms are constructed have a great many features in common.

Because of these cellular and molecular commonalities, scientists have long been confident about transferring subcellular information across taxonomic lines. Because of species differences, as a consequence of adaptation to different biological niches, conclusions of biomedical interest must be verified through tests in mammals, primates, and humans. Generally, the lower a structure or process is on the hierarchical scale of structural complexity (ranging from atoms to organisms), the broader the taxonomic distances over which it can be transferred or extrapolated.

It became clear from the workshops that the common features that exist at relatively high levels of organization are much more extensive than was perceived at the outset. Shared molecular features, which are associated with signalling strategies broadly distributed in the living world, are illustrated by the following relationships, all of which were discussed in the workshops:

- vertebrate peptide hormones, including insulin, and immunochemically similar neurotransmitters present in many invertebrates;

- the existence of estrogen receptors in yeast;

- the close similarity between some coelenterate, molluscan, and mammalian neuropeptides;

- the similarity between endocrine control components in tadpoles and mammals;

- the similarity in the response of sponge cells and human neutrophils to nonsteroidal antiinflammatory drugs; and

- the similarity of developmental genes in insects and vertebrates.

In a series of 14 tables,. Kramer (in press) compiled information on the occurrence and distribution of peptide and peptide-like substances in insects and other nonmammalian life forms. These tables include such subsets of data as effects of vertebrate insulin on invertebrates and the occurrence of insulin-like peptides in insects and other invertebrates, including tissue source, assay, and quantity for 27 organisms. This compilation is exemplary of the many-to-many modelling strategy and would provide a framework for a significant part of the matrix described below. Protein and nucleic acid sequence libraries would, of course, provide additional matrix elements.

In analyzing the workshops, the committee became impressed with the large body of unexplained intertaxonic analogs and molecular similarities.

This suggested that there are undiscovered organizational principles at the cell, tissue, and organismic levels. That is, we seem to be at a point in the history of biology where new generalizations and higher order biological laws are being approached but may be obscured by the simple mass of data and volume of literature. This feeling of impending discovery led the committee to reconsider the importance of organizing this material as a way of extending the modelling concept.

This led back to the notion of "the matrix of biological knowledge" that is the complete data base of published biological experiments structured by the laws, empirical generalizations, and physical foundations of biology and connected by all the interspecific transfers of information. The matrix includes but is more than the computerized data base of biological literature, since the search methods and key words used in gaining access to that base are themselves related to the generalizations and ideas about the structure of biological knowledge.

In referring to the body of ordered biological knowledge as a matrix, we are halfway between the definition of the term as used by the mathematician and by the biologist. The mathematician refers to an ordered array of symbols, the biologist to the substance between structures that holds them together. Biological knowledge of various organisms is held together by evolutionary origins, but is also ordered by phylogenetic and functional requirements.

In the matrix perceived by the committee, the corpus of material is structured by the axes or coordinates that form the conceptual framework. The term matrix usually has a two-dimensional connotation, but the committee has retained the term even though its system has a greater number of dimensions. The notion is at the moment a conceptual rather than a numerical one. Following is a list of possible categories (coordinates) for the structure of the matrix:

- Level of organizational complexity:

  Atoms
  Monomers (amino acids, sugars, nucleotides)
  Metabolic pathways (glycolysis, tricarboxylic acid cycle, carbon dioxide fixation)
  Signal molecules (hormones, pheromones, neurotransmitters)
  Macromolecules (protein, nucleic acids, polysaccharides)
  Organelles (mitochondrion, endoplasmic reticulum)
  Cells (somatic and generative cells)
  Tissues (connective, skeletal, and protective tissues)
  Organs and systems (nervous, circulatory, and reproductive systems)
  Organisms (phylogenetic connections)
  Populations (ecological relationships)

- Phylogenetic status (evolutionary distance from and phylogenetic relationship to _Homo sapiens_ or other taxa of interest)

- Physiological responsiveness, e.g., behavior:
    - Attraction/aversion (chemotaxis in Protista)
    - Memory (short term, long term)
    - Conditioned reflex
    - Numeration
    - Abstraction (problem-solving, tool-making)

- Metabolism. This has already been well structured by metabolic charts and all the energetic and thermodynamic relationships implicit in such charts, e.g., Sallach (1972).

- Regulation. The generality of systems theory, cybernetics, and feed-back analysis is applicable at most hierarchical stages.

- Morphogenesis:
    - Gene expression
        - Repressors
        - Inducers
    - Cell-cell interaction; cell migration
    - Position effects
    - Maternal inheritance

- Molecular structures of physiologically active molecules, including hormones, pheromones, vitamins, peptides, and enzymes.

Other equally valid methods of structuring the matrix should also be considered. These axes or coordinates would allow for the development of a computer-stored data base for the systematic exploration of models once a problem has been defined.

Thus, the body of biological knowledge is constructed of the following major components:

- the observational and experimental data on a very large number of species (most experiments having been conducted on a very small subset of experimentally very favorable systems);

- empirical generalizations and theories based on observations and experiments;

- the biochemical and molecular models of mechanisms based on a reductionist analysis and experiments using basic techniques of physics and chemistry; and

- relevant constructs from applied physical sciences (e.g., geology, meteorology, and hydrology) that describe the interactions of biota with the physical environment.

Biological data are being amassed so rapidly that a properly structured, readily accessible data bank, if it could be developed, would probably be of substantial assistance in leading investigators from a partially analyzed problem to useful models within the matrix. Setting up this computer program would require an ordering that would in itself constitute a theoretical framework for biological science. Some analyses of this type are already part of the various nucleotide and amino acid sequence libraries, but more generalized kinds of knowledge need to be incorporated.

The matrix data base might also include information on projects funded by the NIH, which are already available in computerized form. These state-of-the-art entries would facilitate the exchange of information among investigators using different organisms or working at different levels of biological organization.

The committee recognizes the matrix of biological information as a central concept for modelling. The matrix includes a data bank of experimental findings from which can be derived a set of empirical generalizations. The committee believes that a focus on the matrix could lead modelling and the search for generalizations into better conceptual focus.

The development of the matrix and the extraction of biological generalizations from it are going to require a new kind of scientist, a person familiar enough with the subject being studied to read the literature critically, yet expert enough in information science to be innovative in developing methods of classification and search. This implies the development of a new kind of theory geared explicitly to biology with its particular theory structure (see Schaffner, 1980). It will be tied to the use of computers, which will be required to deal with the vast amount and complexity of the information, but it will be designed to search for general laws and structures that will make general biology much more easily accessible to the biomedical scientist.

The concept of a model as it emerged during the workshops was generalized in the following way. An investigator considers some problem of interest -- a disease process, some normal physiological function, or any other aspect of biology or medicine. The problem is analyzed into its component parts, and for each part and at each level, the matrix of biological knowledge is searched for analogous phenomena. One thus may find different models for various features of the problem. Although it is possible to view the processes involved in interpreting data in the language of one-to-one modelling, the investigator is actually modelling back and forth onto the matrix of biological knowledge. The more coherent and theoretically structured the matrix, the more global is the modelling. Frequently, many of the best connections from the problem to the matrix are made to a single species or subspecies, e.g., Watanabe rabbits. In this case, one-to-one modelling continues to be the most effective strategy, but even this is a special application of many-to-many modelling onto the matrix.

Modelling in the context described above is dynamic and interactive. One goes from the problem at hand to the matrix. With the information obtained, it is possible to conduct experiments that produce data that become part of the matrix, and that enable one to return to the matrix to identify new analogs suggested by the experiment.

In this view of modelling, homology becomes secondary to the dominant role of analogy. Homologous or evolutionary relationships remain, however, of predictive value in finding relevant information within the matrix. The committee has also retained the general view that phylogenetic closeness is accompanied by complexes of relationships of special value in the identification of modelling features from neighboring species.

The body of biological knowledge and the generalizations involve transferability of information within and between the many categories and ways of organizing material within the matrix. Information transfer, for example, can be interhierarchical or intertaxonomic. The first of these is inherent in reductionism, the explanation of phenomena at one level of organization in terms of our understanding at lower levels. There is a further guide to the transfer of information to higher levels in the hierarchy, a teleonomy imposed by evolutionary considerations (Monod, 1971), which allows a certain explanation of phenomena at one level in terms of understanding of the nature of events at a higher level. Homology, associated with evolution, allows intertaxonomic information transfer on the matrix, a feature that can be considerably confused by convergent and divergent evolution.

Because genealogical distance can be expressed quantitatively by degree of DNA hybridization (Sibley and Ahlquist, 1984), the matrix can be structured with anthropocentric bias by considering closeness to *Homo sapiens*. This is probably appropriate for biomedical considerations but may not be optimal for more general (theoretical) biology. It is also clear that the lower the hierarchical level, the more easily information can be transferred intertaxonomically. For example, at the lowest levels, such as atoms and monomers, the mapping relationships are fundamentally the laws of quantum mechanics and organic chemistry. At the levels of molecules and their transformations, the dominant consideration is the ubiquity of metabolic pathways, which arose very early in evolutionary history. With signal molecules as in metabolic pathways, we observe common strategies in different taxa that accomplish certain biochemical and physiological tasks. Le Roith *et al*. (1983) stress how broadly intertaxonomic the signaling strategies, or at least the molecules, are. At the level of the macromolecule, there is the universality of the genetic code and the ubiquity of macromolecular synthesis schemes. At this level, experiments conducted largely on *Escherichia coli* have illuminated basic genetic activities of the entire phylogenetic tree. At the organelle level, the great biological division between procaryotes and eucaryotes manifests itself. Among the eucaryotes, there is remarkable intertaxonomic similarity among the major organelles -- mitochondria, chloroplasts, and membrane systems.

An excellent example of modelling onto the matrix of biological knowledge is found in studies on myasthenia gravis. This is discussed in

Appendix E, but also is reviewed here from a somewhat different perspective to emphasize the structure of this kind of modelling. For many years there was presumptive evidence that the metabolic defect causing the symptoms of this disease was located at the neuromuscular junction. Curare mimicked the disease, and decuratizing agents (anticholinesterases) relieved its symptoms. The postulate was finally made more explicit through research on alpha-bungarotoxin, a component of snake venom that irreversibly binds to acetycholine receptors (AChRs). By using the toxin, AChR from electric eels and electric rays could be purified. Injection of the purified AChR into rabbits to make antibodies to further characterize and localize the receptor produced myasthenic symptoms in the rabbits. This finding was unexpected, and sufficient conservation of protein structure between fish and rabbit to result in immunologic cross reactivity had not been predicted. Subsequently, experiments with alpha-bungarotoxin indicated that myasthenic patients had a deficit of AChRs due to increased rates of degradation. The methodology for these studies was developed in studies on chicken muscles in culture, which had been conducted with the goal of establishing turnover rates for the receptor protein.

These studies are now leading to an understanding of the immunological and molecular nature of the disease and to a rational and effective theory of its treatment. Data have emerged from studies on the electric eel, electric ray, cobra, chicken, mouse, rabbit, rat, human, and various tissue and cell culture systems. Progress in understanding and treating myasthenia gravis depends critically on the ability to model onto the entire matrix as well as on the utilization of information on action potentials and neurotransmission, which came to the matrix from studies on nonmammalian vertebrates and on squid and other invertebrates. Moreover, much of the critical work was done as basic research, and several findings, such as the experimental induction of myasthenia in rabbits, were quite fortuitous.

The study of myasthenia gravis shows how a given biomedical problem can often be broadly related to the data base in the matrix, thereby identifying numerous models for the processes or structures under study. As a result of this highly interrelated matrix structure, every valid biological study is potentially of biomedical value, since it is difficult, if not impossible, to predict where in the matrix a given study is going to make connections. Resource allocation should be focused on good work and the intuition of good scientists in sensing which taxa and areas of study will be most useful in developing the total matrix. The interest of the community of biologists in a given area of work generally depends on the extent to which that work connects with other areas of the matrix.

Thus, biology involves something less than an axiomatic formulation (Schaffner, 1980) and something far more than a simple collection of data. The interrelatedness of all living things provides a highly structured but nonquantitative theory. In this context, data on any organism form part of the matrix of biomedical knowledge.

It follows that any organism or biological preparation that has paid off in data transferable to humans should be studied further. This may be illustrated by examples from the workshop on models for the study of development (see Appendix E). In that workshop, Gerhart noted that, although the ascidian egg serves as an excellent model system in the study of early embryonic development, its payback as a model is limited by information gaps concerning the organism's life cycle, longevity, reproductive cycle, genetics, and laboratory cultivation. Similarly, Wood suggested that the utility of mapping cell fates in the nematode *Caenorhabditis elegans* will be even greater when the genes controlling cell fate are identified.

One type of observation kept recurring in all the workshops. A small number of species have achieved prominence as research tools because they have been very extensively studied from a number of perspectives and thus provide well-understood paradigms that have been described in detail in terms of genetics, biochemistry, physiology, and other aspects. They are called high connectivity models. Two examples are *E. coli* and *Drosophila melanogaster*. It is clear that such concentration on many aspects of the biology of a single organism is an effective way to build connectivity within the matrix and should thus be supported. On the other hand, taxonomic breadth is also required, since species that may lend themselves particularly well to the study of specific problems often cannot be predicted. The successful search for model systems needed in the future will require a broad-based systematic biology and comparative physiology. Concentration on selected species and taxonomic diversity are not mutually exclusive. They both take part in the establishment of a maximally useful matrix of biological knowledge.

REFERENCES

Kramer, K. In press. Vertebrate hormones in insects. Pp. 511-536 in G. Kerkut and L. I. Gilbert, eds. Comprehensive Insect Physiology, Biochemistry and Pharmacology. Embryogenesis and Reproduction, Vol. 7. Pergamon Press, New York.

LeRoith, D., J. Shiloach, M. Berelowitz, L. A. Frohman, A. S. Liotta, D. T. Krieger, and J. Roth. 1983. Are messenger molecules in microbes the ancestors of vertebrate hormones and tissue factors? Fed. Proc. 42:2602-2607.

Monod, J. 1971. Chance and Necessity. Knopf, New York. 198 pp.

Sallach, H. J. 1972. Intermediary Metabolism Charts. P. L. Biochemicals, Inc., Milwaukee, Wisconsin.

Schaffner, K. F., 1980. Theory structure in the biomedical sciences. J. Med. Philos. 5:57-97.

Sibley, C. G., and J. E. Ahlquist. 1984. The phylogeny of hominoid primates as indicated by DNA-DNA hybridization. J. Mol. Evol. 20:2-15.

6

Conclusions and Recommendations

CONCLUSIONS

1. There are two principal types of models in biomedical research, both based on analogy. The first type seeks similarities between a process or structure within an organism of interest and other organisms, or parts of organisms, where analogous processes or structures occur. The second type models processes or structures with conceptual, mathematical, or mechanical analogs. Models are used because they possess a simpler or more accessible structure or mechanism in comparison to the object of primary interest, or because certain classes of experiments cannot be carried out on humans.

2. Models of the first type, i.e., organisms with a characteristic of interest, also fall into two categories. The traditional one-to-one model seeks a surrogate that contains an entire complex of features that are analogous to the process or disease of interest in the primary organism, which is usually *Homo sapiens*. The emerging many-to-many concept of modelling begins by analyzing a process or disease into its component parts and then finding and utilizing for each component analogous behavior in many taxa throughout the living world.

3. In a broad context, biomedical research contributes to a great matrix of phylogenetically and biochemically interconnected information about living organisms, including humans. Because of common evolution and shared environments, the biology of human health and disease is inextricably interconnected with an understanding of all other organisms. Knowledge of human health and disease increases as we discern more and more of the interconnections, interrelationships, and reciprocal relationships that link human beings to the rest of the living world. In this context, the nature of modelling can be visualized as many-to-many modelling to the matrix of biological knowledge. Here, the value of the system studied resides in its contribution to the discovery of significant interconnections in the matrix. This may be perceived as indirect modelling to the human species, since contributions to the matrix all ultimately contribute to fuller understanding of the human system.

4. The selection of a system for research should be based on the following considerations:

- appropriateness as an analog; transferability of information
- genetic uniformity of organisms, where applicable
- background knowledge of biological properties
- cost and availability
- generalizability of the results
- ease of experimental manipulation
- ecological consequences and ethical implications

5. The ultimate objective of modelling in medical research is generally to understand the human organism. Processes or systems being modelled may range from the molecular to the organismal level and may include both healthy and disease states. To be successful in terms of yielding information and interpretation that permits prediction, any model must be an effective analog (i.e., it must reflect an agreement, likeness, or correspondence in function) to the object or process being modelled. In biomedical research, the common evolutionary origin of all organisms and their consequent shared portions of genomes makes homology a useful tool in seeking analog models.

Because phylogeny and evolution are not understood in all details, however, homology cannot be uncritically applied to the search for good biomedical models and model systems. Divergent evolution may result in some homologs that are bad analogs, whereas convergent evolution may lead to analogy where little or no homology is apparent or expected.

6. To the degree that any biological activity can be described quantitatively, it can be expressed with a mathematical model. Such models cannot be homologs.

7. Biological models or model systems derived from or consisting of nonmammalian organisms, or cell and tissue culture systems derived from vertebrates, can reduce the use of mammals, especially in the early stages of some investigations.

8. Vertebrates, especially mammals, provide essential one-to-one models for many specific human disease processes. Since understanding the biology of the human being in good health and in various disease states is the overall mission of the National Institutes of Health (NIH), it is easy to see the human as the ultimate object of modelling in biomedical research. Models, then, would be chosen with reference to human disease states or normal human function, i.e., one-to-one modelling. In the past, this strategy has paid great dividends in understanding and controlling disease states as well as health, and it is expected to continue as an important strategy in biomedical research.

9. Organisms and preparations are often initially studied for reasons quite different from those that subsequently give them value as models. It would thus be inimical to progress in biomedical research to stipulate too narrowly how model development should proceed. Although systematic assessment may permit us to identify important gaps in our knowledge, it is not possible to stipulate particular taxa, or cell or tissue culture systems, as being more desirable sources of knowledge than others. The systems that have provided the most important contributions have often been those that have been interesting to investigators and studied in their own right.

Historically, the diversity of organisms and of preparations derived from them has contributed knowledge of great benefit to medicine as well as to biology in general. In the past, it has been difficult to predict accurately which organisms or preparations would yield the most useful insights and information for medicine or biology. Basic research on many kinds of organisms holds the greatest promise of producing meaningful contributions to the matrix of biological knowledge.

10. Vertebrates other than mammals, including birds, reptiles, amphibians, and fish, provide good experimental material for investigation of many biological processes. All have contributed significantly to the understanding of development and regulation in higher vertebrates.

11. Microorganisms of many kinds have served effectively in a variety of biomedical studies. The highly conserved molecular genetic machinery is essentially the same for all living organisms. Whereas bacteria have been used in cellular biochemistry, especially in metabolism studies, yeast and fungi have served in the study of eucaryotic genetics, nutrition, and metabolism. Protozoa and other eucaryotic microorganisms have contributed to our understanding of genetics, cell aggregation, differentiation, and cellular fine structure.

12. Often, a human process or function must be studied in a model simpler than an intact vertebrate. In some of these cases, cell and tissue culture systems have contributed to the solution of many biomedical problems. For example, major recent advances in our knowledge of the immune system made possible by cell cultures would have been virtually impossible to achieve in intact vertebrates. Cell culture techniques, including monoclonal culture, have also been responsible for the elucidation of various mechanisms, such as differentiated cell function, regulation of cell proliferation, cell-cell and cell-substratum interactions, and carcinogenesis.

In vitro systems are not limited to cell and tissue cultures. Intact tissues as well as tissue and organ parts may often be maintained for sufficient time to permit their use as models for functions of the intact organism from which they were extirpated.

13. For several aspects of human biology, mammals provide the best, and in some cases the only, biological models. In some instances, primates are the only animals that can serve as models for a specific purpose. Even when models can be derived from cultured mammalian materials or from mathematics, it is necessary to verify results and conclusions by comparing them with experimental data derived from intact mammals.

14. Mathematical modelling is a useful investigational strategy when closely coupled to experimentation. Opportunities for mathematical modelling are present in all areas of biomedical research and at all levels of biological organization. Often, as for example in enzyme kinetics or in population growth, mathematical formulations become so thoroughly integrated into the matrix of biological knowledge that they cease to be recognized as models. Regrettably, there is often a gap in communication and understanding between mathematical modelers and experimental investigators.

15. Chemical inhibitors of biological functions and processes have served as valuable probes of normal function, especially at the cellular and subcellular levels. If more chemicals of environmental concern were tested for their inhibitory effects in a greater array of organisms, the comparative data that would be generated would be mutually enriching for toxicology and for general comparative biology.

16. Considering the great strides in our understanding of biology and medicine that have resulted from the study of microorganisms, invertebrates, and lower vertebrates, the proportion of NIH resources that supports research in this area may be small in comparison to the resources dedicated to research with mammals.

17. Biomedical research should not be regarded separately from biological research. Applications of basic biology to clinical medicine are often apparent, but they cannot, in general, always be predicted. This consideration justifies NIH support for basic biology. The search for new and improved models cannot be programmatically directed, because comparative biology provides insufficient guidelines. Experience indicates that the information yield from basic research throughout the general matrix of biological knowledge will be substantial, in terms of increased knowledge of human function.

RECOMMENDATIONS

1. Proposals for the study of invertebrates, lower vertebrates, microorganisms, cell and tissue culture systems, or mathematical approaches should be regarded as having the same potential relevance to biomedical research as proposals for work on systems that are phylogenetically more closely related to humans. Support should be given to good research without taxonomic or phylogenetic bias on the part of the sponsor and should include comparative and phylogenetic studies.

2. As favorable systems are identified, the NIH should strive to make them readily available to the research community. Following are some examples of how this could be done:

- provision of support to supply organisms for research
- maintenance of stock centers for mutant strains and for cell lines
- facilitation of access to computer programs for biomedical modelling
- maintenance of data bases like those for protein and DNA sequences
- provision of long-term support for collections of cloned genes and useful vectors or collections of monoclonal antibodies

3. For many biomedical problems, there are no adequate nonmammalian models. In such cases, the NIH should continue to support research using the best mammalian models. Efforts to find additional models for normal function and for specific diseases should be continued, although adequate alternatives to the use of mammals will not always be found. This will be especially true for problems that involve complex systems such as higher nervous function. The assumption that adequate alternatives to mammals can always be found is incorrect and should not influence the distribution of research resources. Moreover, when elucidation of mammalian biology is the objective, results from nonmammalian organisms must be validated in mammals.

4. Although the development of techniques for the use of novel experimental organisms probably proceeds most efficiently when mandated by a specific ongoing line of investigation, the NIH should consider supporting proposals whose objective is the development of model systems for specific research areas. Some mechanism other than the current ad hoc process should be established for the evaluation of proposals that do not fall conveniently into existing study sections. Indeed funds might be targeted for the development of new model systems that appear to be particularly promising.

Because models are often identified not through a planned search but rather as a consequence of interest in an organism for its own sake, the NIH should encourage meritorious studies regardless of whether direct biomedical application is apparent or implied. Particular attention should be given to little known taxa, which may be untapped sources of valuable organisms for biomedical research.

5. The NIH should explore the possibility of creating a clearinghouse to encourage the use of nonmammalian model systems for testing the effects of exposures to chemicals of interest to environmental toxicologists. Such a clearinghouse function might, for example,

encourage NIH-supported researchers studying nonmammalian systems to test chemicals of potential concern as identified by the National Toxicology Program.

6. The NIH should consider encouraging interest in nonmammalian systems through postdoctoral fellowships, symposia, and direct support of model development. Development of nonmammalian systems might also be encouraged by providing to investigators in computer-accessible form elements of the matrix of biological knowledge. Direct support of model development should include particular attention to potential high-connectivity models.

7. Although the NIH should encourage the community of biomedical researchers to think critically and creatively about the development of new and improved models, particularly for human diseases, the selection of a system or organism for proposed research is best left to the individual investigator.

8. The matrix of biological knowledge should be further investigated as a potential tool in biomedical research under the aegis of the NIH. The concept must be sharpened and subjected to test as to its utility.

Appendices

These summaries are based largely on abstracts of presentations prepared by workshop participants. Because there was insufficient time to permit review of the summaries by the workshop participants, the committee takes responsibility for any errors in presentation, interpretation, or citation.

Discussion of particular organisms is not intended to imply their superiority over other systems but to illustrate the considerations that are relevant to the selection and evaluation of model systems.

Appendix A

Nobel Prize Winners In Physiology or Medicine, from 1901 through 1984[a]

Year	Winner	Experimental Organism	Contribution
1901	Emil von Behring	Guinea pig	Diphtheria serum therapy
1902	Ronald Ross	Mosquito, human, crow, pigeon, lark, sparrow	How malaria enters the organism and methods for combating it
1903	Niels Finsen	Bacteria (various)	Treatment of diseases, especially lupus vulgaris, with concentrated light radiation
1904	Ivan Pavlov	Dog	Physiology of digestion
1905	Robert Koch	Cattle, sheep	Investigations and discoveries in relation to tuberculosis
1906	Camillo Golgi	Human, horse, dog	Nervous system structure
	Santiago Ramon y Cajal	Human, birds, reptiles, mammals	
1907	Charles Laveran	Human, protozoa, mammals, birds, fish	Role of protozoa in causing diseases
1908	Elie Metchnikov	Starfish, birds, fish	Immunity
	Paul Ehrlich	Guinea pig	
1909	Theodor Kocher	Human	Physiology, pathology, and surgery of the thyroid gland

---

[a] Compiled from Sourkes (1966), the Nobel Foundation Calendar (1977), The New Encyclopedia Britannica (1984), and news items from Science (1977-1984).

Year	Winner	Experimental Organism	Contribution
1910	Albrecht Kossel	Bird	Knowledge of cell chemistry through work on proteins, including nucleic substances
1911	Allvar Gullstrand	Human	Dioptrics of the eye
1912	Alexis Carrel	Dog	Vascular suture and grafting of blood vessels and organs
1913	Charles Richet	Dog, rabbit	Anaphylaxis
1914	Robert Barany	Human	Physiology and pathology of the vestibular apparatus
1915-1918	Reserved[b]		
1919	Jules Bordet	Guinea pig, horse, rabbit	Work on immunity
1920	August Krogh	Frog	Discovery of capillary motor-regulating mechanism
1921	Reserved		
1922	Archibald Hill	Frog	Production of heat in the muscle
	Otto Meyerhof	Frog	Relationship between consumption of oxygen and metabolism of lactic acid in muscle
1923	Frederick Banting	Dog, rabbit	Discovery of insulin
	John Macleod	Dog, rabbit, fish	
1924	Willem Einthoven	Dog, human, frog	Mechanism of the electrocardiogram
1925	Reserved		
1926	Johannes Fibiger	Rat, cockroach	Discovery of *Spiroptera* carcinoma

b No prize given.

Year	Winner	Experimental Organism	Contribution
1927	Julius Wagner-Jauregg	Human	Malaria inoculation in the treatment of dementia paralytica
1928	Charles Nicolle	Monkey, chimpanzee, pig, rat, mouse	Work on typhus
1929	Christiaan Eijkman	Chicken	Antineuritic vitamin
	Frederick Hopkins	Bird	Growth-promoting vitamins
1930	Karl Landsteiner	Human	Human blood groups
1931	Otto Warburg	Yeast	Nature and mode of action of the respiratory enzyme
1932	Charles Sherrington	Frog, dog, cat	Function of neurons
	Edgar Adrian	Frog	
1933	Thomas Morgan	Vinegar fly, fruit fly (*Drosophila*)	Role of the chromosome in heredity
1934	George Whipple	Dog	Liver therapy in anemia
	George Minot	Human	
	William Murphy	Human	
1935	Hans Spemann	Amphibian	Organizer effect in embryonic development
1936	Henry Dale	Cat, mammals, frogs, reptiles, birds	Chemical transmission of nerve impulses
	Otto Loewi	Frog	
1937	Albert Szent-Gyorgyi	Potato, pepper, lemon	Biological combustion processes, with special reference to vitamin C and the catalysis of fumaric acid

Year	Winner	Experimental Organism	Contribution
1938	Corneille Heymans	Dog	Role of the sinus and aortic mechanisms in regulation of respiration
1939	Gerhard Domagk	Mouse, rabbit	Antibacterial effects of Prontosil
1940-1942	Reserved		
1943	Henrik Dam	Rat, mouse, dog, chick	Discovery of vitamin K
	Edward Doisy	Alfalfa, fish meal	
1944	Joseph Erlanger	Bullfrog	Differentiated functions of single nerve fibers
	Herbert S. Gasser	Cat	
1945	Alexander Fleming	*Penicillium* and various bacteria	Penicillin and its curative effect in various infectious diseases
	Ernst Chain	*Penicillium* and various bacteria	
	Howard Florey	*Penicillium* bacteria (various)	
1946	Hermann Muller	Fruit fly (*Drosophila*)	Production of mutations by x-irradiation
1947	Carl Cori	Frog	Catalytic conversion of glycogen
	Gerty Cori	Frog	
	Bernardo Houssay	Toad, dog	Role of the pituitary hormone in the metabolism of sugar
1948	Paul Muller	Beetle, fly, aphid, gnat	Efficiency of DDT as a contact poison against several arthropods

Year	Winner	Experimental Organism	Contribution
1949	Walter Hess	Cat	Functional organization of the interbrain as a coordinator of the activities of the internal organs
	Antonio Moniz	Human	Therapeutic value of leucotomy in certain psychoses
1950	Edward Kendall	Cattle	Hormones of the adrenal cortex -- structure and biological effects
	Tadeuz Reichstein	Dog	
	Philip Hench	Human	
1951	Max Theiler	Monkey, mouse	Vaccine against yellow fever
1952	Selman Waksman	Guinea pig	Streptomycin, the first antibiotic effective against tuberculosis
1953	Hans Krebs	Pigeon	Citric acid cycle
	Fritz Lipmann	Pigeon	Coenzyme A and its role in intermediary metabolism
1954	John Enders	Monkey, human, mouse, poliovirus	Cultivation of poliomyelitis viruses in various types of tissue cultures
	Thomas Weller	Monkey, human, mouse, poliovirus	
	Frederick Robbins	Monkey, human, mouse, poliovirus	
1955	Hugo Theorell	Horse	Nature and mode of action of oxidative enzymes
1956	André Courand	Human	Heart catheterization and pathological changes in the circulatory system
	Werner Forssman	Human	
	Dickinson Richards, Jr.	Human	

Year	Winner	Experimental Organism	Contribution
1957	Daniel Bovet	Dog, rabbit	Production of synthetic curare and its action on the vascular system and skeletal muscles
1958	George Beadle	Fruit fly (*Drosophila*), snapdragon	Gene regulation of definite chemical events
	Edward Tatum	*Neurospora*	
	Joshua Lederberg	*Escherichia coli*	Genetic recombination and organization of the genetic material of bacteria
1959	Severo Ochoa	*Azotobacter*, *E. coli*, snake	Mechanisms in the biological synthesis of RNA and DNA
	Arthur Kornberg	*E. coli*, *Mycobacterium*, bacteriophage T_2, calf	
1960	Macfarlane Burnet	Rabbit	Acquired immunological tolerance
	Peter Medawar	Cattle, mouse, sheep, chick	
1961	Georg von Bekesy	Guinea pig	Physical mechanism of stimulation within the cochlea
1962	Maurice Wilkins	Tobacco mosaic virus	Molecular structure of nucleic acids and its significance for information transfer in living material
	James Watson	Turnip yellow mosaic virus, tobacco mosaic virus, *E. coli*	
	Francis Crick	Turnip yellow mosaic virus, tobacco mosaic virus, *E. coli*	

Year	Winner	Experimental Organism	Contribution
1963	John Eccles	Cat, frog	Ionic involvement in excitation and inhibition in the peripheral and central portions of the nerve
	Alan Hodgkin	Squid, crab	
	Andrew Huxley	Squid, crab	
1964	Konrad Bloch	Rat	Mechanism and regulation of cholesterol and fatty acid metabolism
	Feodor Lynen	Yeast	
1965	François Jacob	Bacteriophage, *E. coli*	Genetic control of enzyme and virus synthesis
	Andre Lwoff	Bacteriophage, *E. coli*	
	Jacques Monod	Bacteriophage, *E. coli*	
1966	Peyton Rous	Rat, hen, rabbit, virus	Tumor-inducing viruses
	Charles Huggins	Dog, rat	Hormonal treatment of prostatic cancer
1967	Ragnar Granit	Frog, rabbit, snake, eel, crab	Primary physiological and chemical visual process in the eye
	Keffer Hartline	Frog, rabbit, snake, eel, crab	
	George Wald	Fish, squid, chicken	
1968	Robert Holley	Rat, yeast	Interpretation of the genetic code and its function in protein synthesis
	Har Gobind Khorana	Rat, *E. coli*	
	Marshall Nirenberg	*E. coli*, yeast, tobacco mosaic virus, frog (*Xenopus*), guinea pig	

Year	Winner	Experimental Organism	Contribution
1969	Max Delbruck	Bacteria, bacteriophage (various)	Replication mechanism and the genetic structure of viruses
	Alfred Hershey	Bacteria, bacteriophage (various)	
	Salvador Luria	Bacteria, bacteriophage (various)	
1970	Bernard Katz	Squid, frog	Transmitters in the nerve terminals and the mechanism for their storage, release, and inactivation
	Ulf von Euler	Cat	
	Julius Axelrod	Rat	
1971	Earl Sutherland, Jr.	Mammalian liver	Mechanisms of the action of hormones
1972	Gerald Edelman	Guinea pig, human	Chemical structure of antibodies
	Rodney Porter	Rabbit, human	
1973	Karl von Frisch	Bee	Organization of individual and social behavior patterns in animals
	Konrad Lorenz	Bird	
	Nikolaas Tinbergen	Bird	
1974	Albert Claude	Chicken, guinea pig, rat	Structural and functional organization of cells
	Christian de Duve	Rat	
	George Palade	Guinea pig	
1975	David Baltimore	Poliovirus, monkey	Interaction between tumor viruses and the genetic material of the cell
	Renato Dulbecco	Equine virus, simian virus polyoma virus, DNA tumor virus, mouse	
	Howard Temin	Rous sarcoma virus, chicken	

Year	Winner	Experimental Organism	Contribution
1976	Baruch Blumberg	Human	New mechanisms for the origin and dissemination of infectious diseases
	Carleton Gajdusek	Human, chimpanzee	
1977	Roger Guillemin	Sheep, swine	Hypothalamic hormones
	Andrew Schally	Sheep, swine	
	Rosalyn Yalow	Human	Development of radioimmunoassay (RIA), and of the principles underlying RIA
1978	Werner Arber	Bacteria	Discovery and application of restriction endonucleases
	Hamilton Smith	Bacteria	
	Daniel Nathans	Bacteria	
1979	Allan M. Cormack	Human	Development of the x-ray diagnostic technique, computer-assisted tomography (CAT)
	Godfrey Hounsfield	Human, pig	
1980	Baruj Benacerraf	Mouse, guinea pig	Identification of histocompatibility antigens and elucidation of their action
	Jean Dausset	Human	
	George Snell	Mouse	

Year	Winner	Experimental Organism	Contribution
1981	Roger Sperry	Human	Functions of the cerebral hemispheres
	David Hubell	Cat, monkey, human	Processing of visual information by the brain
	Torsten Wiesel	Cat, monkey, human	
1982	Sune Bergstrom	Soybean, ram	Biochemistry and physiology of prostaglandins
	Bengt Samuelsson	Rabbit	
	John Vane	Guinea pig, rabbit	
1983	Barbara McClintock	Corn	Discovery of mobile genetic elements
1984	Cesar Milstein	Plasma cells	Development of technique for monoclonal antibody formation
	Georges Kohler	Plasma cells	
	Neils Jerne	Theory	

REFERENCES

The New Encyclopedia Britannica. 1984. Nobel Prize Winners. Pp. 371-372 in Micropaedia, Vol. VII. Encyclopedia Britannica, Chicago.

Nobel Foundation Calendar (1977-78). 1977. Sturetryckeriet AB, Stockholm, Sweden. 138 pp.

Science. 1977-1984. Various articles on Nobel Prize winners. 198:594-596; 202:1069-1071; 206:1060-1062; 210:621-623; 214:518-520; 218:765-768; 222:402-405; 226:1025-1028.

Sourkes, T. S. 1966. Nobel Prize Winners in Medicine and Physiology 1901-1965. Abelard-Schuman, New York. 464 pp.

Appendix B

Model Systems in Cellular Immunology

Workshop Program

April 27, 1984
Washington, D.C.
Gordon Sato, Organizer
John Farrar, Chairman

Gordon Sato	Introductory Talk: A Cultural Approach to Cellular Endocrinology
Henry N. Claman University of Colorado Medical School Denver, Colo.	*In Vivo* and *In Vitro* Methods for the Measurement of Delayed Hypersensitivity
John Farrar Hoffman-LaRoche, Inc. Nutley, N.J.	Advances in Model Systems Useful in the Development of Immunomodulatory Compounds
Joost Oppenheim National Cancer Institute Frederick, Md.	Advances and Limitations in the Use of *In Vitro* Models of Cytokine Function
Ellis Reinherz Dana Farber Cancer Research Institute Boston, Mass.	Human T Cell Clones and Monoclonal Antibodies as Probes to Delineate the T Cell Antigen Receptor
Fritz Bach University of Minnesota Minneapolis, Minn.	Advances in the Cloning and Manipulation of Human T Cells
Donald Mosier Fox Chase Institute for Cancer Research Philadelphia, Pa.	Problems and Potentials for *In Vitro* Models of Antibody Synthesis
Verner Paetkau University of Alberta Edmonton, Alberta, Canada	Strategies for Structure-Activity Analysis of Lymphokines
Paul Nadler Hoffman-LaRoche, Inc. Nutley, N.J.	Relevance of *In Vitro* Immunologic Responses to Clinical Studies with Immunomodulators

Workshop Summary

Dramatic advances in understanding the workings of the immune system have been made thanks to techniques that permit the study of the immunological properties and behavior of various mammalian cell types in culture. Isolated from the tissue and organ level effects, which characterize the immune response of the intact organism, the exclusively cellular attributes of the system can be analyzed. Because *in vitro* methodologies have been so effectively exploited in immunology, this approach was selected for in-depth analysis in the first workshop of the series.

Henry N. Claman (University of Colorado Medical School) discussed methods for studying delayed hypersensitivity responses in mice. He emphasized that his model stressed *in vivo* assays, which should be correlated with *in vitro* methods. This complex biological response involves activation of a variety of cell types. In mice, the response is induced by painting the animals' abdomens with 2,4-dinintrofluorobenzene, followed by a challenge on the ears with the same antigen and then measuring the resulting ear swelling (Phanuphak *et al.*, 1974a). The response is not accompanied by the production of antibodies to the chemical used for induction. It is dependent on the presence of T cells, since it does not occur in mice lacking such lymphocytes. This treatment results in the induction of at least three distinct T cell types *in vivo*, only one of which is believed to be responsible for the delayed hypersensitivity reaction (Claman *et al.*, 1980). Proliferation of T cells is not a good indicator of the delayed hypersensitivity response in all respects.

The *in vivo* model can be used to study specific immune suppression, which is often produced by the same agent used to produce sensitization. Injection of dinitrofluorobenzene in mice not only suppresses the sensitization produced by painting the substance on the animals' abdomen but also results in the production of T suppressor cells (Germain and Benacerraf, 1980; Phanuphak *et al.*, 1974b). Activation of suppressor T cells requires a specific antigen, but the final suppressive event is not antigen-specific. Thus T cell circuits are specific in how they are turned on but nonspecific in their suppression. Analysis of these phenomena with *in vivo* models has progressed as far as possible. *In vitro* models are now needed to define and clarify the mechanisms of this response. This is challenging, since cell recruitment and migration between organs is a very important part of the response of an intact immune system yet can obviously not be duplicated by present *in vitro* methods. Corticosteroids are known to alter cell traffic (migration), which accounts for their activity as antiinflammatory agents (Levine and

Claman, 1970). In summary, it is necessary to adopt a reductionist approach through the use of *in vitro* systems to evaluate components of the delayed type hypersensitivity response followed by validation in *in vivo* experimentation.

The use of *in vitro* model systems in the development of immunomodulatory drugs was discussed by John Farrar (Hoffman-LaRoche, Inc.). He pointed out that numerous specific points in the immune system could serve as targets for pharmacological modulation and that *in vitro* technologies have provided model systems, which can be used in detailed studies of cellular responses. Such models are in use as rapid screens of immunomodulatory compounds, which undergo final evaluation as potential drugs in whole animals tests. In these instances, the *in vitro* model systems have been found to be extremely faithful to the *in vivo* tests.

Joost Oppenheim (National Cancer Institute) talked about advances and limitations in the use of *in vitro* models for investigating cytokine factors. Monocytes or macrophages are cells that respond to irritants by ingesting and degrading them (Metchnikoff, 1905). Pieces of these digested materials are then exported to the outer surface of the macrophage cell membrane where they act on T cells, stimulating them to proliferate (Ziegler and Unanue, 1982). Proliferating T cells in turn produce substances (lymphokines) that act on other T cells and ß cells, augmenting immune response including the induction of ß cells to produce antibodies against the irritant (Oppenheim *et al*., 1981). T cells also induce other effects on macrophages. Since this is a very complex system to study *in vivo*, it has therefore been isolated for *in vitro* analysis. Cytokines and interleukins were discovered in this manner. These factors have been studied in tissue culture. Cell cultures serve both as sources of cytokines and assay systems for their detection.

Studies based on these approaches have shown that cytokines are intracellular messengers produced by many cell types, which can act on other cells in the immune system. The substances themselves are polypeptides and glycopeptides that are active at concentrations as low as one part in 100 million (for review see Farrar *et al*., 1982). They are synthesized and secreted at the site of action and might be very difficult to isolate *in vivo*, except in situations of overproduction due to inflammation or aberrant production. A variety of *in vitro* assays are now available for determining the presence of these substances. These include, for example, the stimulation of macrophages to migrate or ingest microorganisms. Similarly, lymphokines can be used to activate T cells to help ß cells produce antibody or lyse target cells. Advantages of the *in vitro* systems in these studies are obvious. Disadvantages and limitations to this approach include nonrenewable materials, loss of normal tissue architecture and cellular matrix, and possible contamination of the cultured cells by external agents or their own waste products. These assays are semiquantitative and must rely on the use of whole animals as a source of the cells used.

T cell clones and monoclonal antibodies are used as probes to delineate T cell antigen receptors. These were discussed by Ellis Reinherz (Dana Farber Cancer Research Institute), who works with human T cell cultures. There are many sets of T cells, each of which has unique cell surface characteristics. Through the use of an *in vitro* system, one can learn about cytotoxic T cells in humans. What are the rules governing their function? What are the cell functions? What are the cell surface markers? *In vitro* studies have shown the evolution of separate populations of T cells that recognize separate antigens on the surface of the same target cells: T-8 cells recognized HLA antigen, whereas T-4 cells recognized the IA antigen (Meuer *et al*., 1982). It was thus discovered that several cell surface molecules are involved in T cell, target cell recognition. Reinherz has been able to develop antibodies to specific individual human T cell types (Acuto *et al*., 1983). The biochemical mechanisms involved in stimulating T cell proliferation are under study in this *in vitro* model system (Meuer *et al*., 1984). The insights into the immune response provided by these studies have relevance to diagnosis and treatment of autoimmune diseases and T cell receptors. This is difficult to do *in vivo*, and is an instance where an understanding of a critical biological function is best approached through use of an *in vitro* model.

In discussing advances in the cloning and manipulation of human T cells, Fritz Bach (University of Minnesota) pointed out that there is a vast amount of literature on the use of *in vitro* models in studies of immunology and that there is no need to defend their use in cell biology. How mechanisms uncovered with *in vitro* systems relate to conditions *in vivo* is a matter of faith, since the direct connections between them are often fragile. The widest conceptual gap occurs when transferring information from simple models to biological functions within the whole animal. Bach's studies are directed at understanding the biological basis for transplant rejection. He uses a mixed leucocyte culture as a model for the allograft reaction and for predicting kidney graft survival. He examines both T helper lymphocytes and T cytotoxic lymphoctes. Using this *in vitro* model, Bach found evidence that cytotoxic cells can be divided into two classes: (1) the classical cytotoxic cell that is not stimulated to proliferate by the antigen for which it is specific and whose proliferation was dependent on the addition of exogenous help and (2) the cytotoxic cell that responds proliferatively to helper-independent antigen. He finds cytotoxic cells specific for all classes of antigens. The helper-independent cytotoxic cells, discovered *in vitro*, were also shown to exist *in vivo* -- an observation supporting the fidelity of this *in vitro* model. Intradermal injection of these cells produced delayed hypersensitivity in recipient animals. Through development of specific cDNA probes, it may be possible to obtain markers for different important cells of the immune system. These markers will allow early determination of prospective transplant rejection by a potential recipient. Despite the power of this *in vitro* model to uncover aspects of mechanisms of immune system function not observable in whole animals, the proof of the reality of these findings was made possible by demonstrating that these cells and their special functions occurred in vivo. Observations made initially in vitro are validated through demonstration of their existence in vivo.

Donald Mosier (Fox Chase Institute for Cancer Research) discussed the problems and potentials of in vitro models of antibody synthesis. He stressed the point that in vitro systems for antibody synthesis are extremely useful models for studying cellular differentiation. Activation of resting T and ß cells and a whole host of other biological functions are observable within 4 to 5 days in vitro. It is an elegant system, but it is not always clear that these in vitro systems are totally reflective of their in vivo counterparts. However, it is clear that the study of in vitro antibody responses has led to a major portion of our understanding of immune system responses and is totally responsible for our knowledge of lymphokine-dependent cellular interactions (Mosier and Subbarao, 1982). Such systems have allowed the study of ß-cell differentiation all the way to the production of antibody. Using an in vitro system, one can make 200 to 400 cultures from a single mouse. If these same studies were to be conducted in vivo, they would require 200 to 400 mice to achieve the same number of observations. There are limitations to the in vitro models, however. Artifacts may occur, and the physiological significance of any finding needs confirmation in an intact animal. Much of our understanding of cellular immunology has yet to be transferred to applications in human biology.

Verner Paetkau (University of Alberta) talked about analysis of structure-activity relationships among the lymphokines. In these investigations, in vitro studies are crucial, since injection of interleukin-2 into an animal results in immediate clearance of the substance, allowing one to see biological effects only if massive doses are given. Using in vitro techniques, Paetkau is attempting to translate the message for synthesis of these substances. The lymphokines are highly active, functioning at the picomole level, and are extremely hydrophobic. He considers these substances to be more like neurotransmitters than hormones and refers to them as "lymphotransmitters." By developing cDNA probes to the mRNA producing these materials, one can assess their biological significance and function in vivo.

Paul Nadler (Hoffman-La Roche, Inc.) described efforts to apply information from in vitro studies of immunomodulatory responses to uses of these same agents in the clinical treatment of immunodeficiency diseases. Drugs enhancing cytolytic T-cell responses in vitro have not yet been successful in treating patients with acquired immune deficiency syndrome (AIDS) (Rook et al., 1983). The response seen in vitro has not yet been repoduced in vivo. Treatments with both alpha and gamma interferons, or with interleukin-2, all of which enhanced killer cell function in vitro, have thus far failed to produce any such measurable effects in AIDS patients. Efforts are under way to produce better animal and in vitro models of this condition.

The extremes in views toward culture systems as models were presented by Fritz Bach (University of Minnesota) and Henry Claman (University of Colorado Medical School). Bach has shown that an exceptional T cell discovered *in vitro* actually existed *in vivo*. He is therefore an enthusiastic exponent of cell culture model systems. Claman, on the other hand, deals with a very complex phenomenon -- delayed hypersensitivity -- and expressed severe doubts as to the applicability of culture systems to this phenomenon.

These differences of opinion date from Harrison (1907), who described the outgrowth of neurites from frog ganglia in culture. Many believed that animals could be taken apart cell by cell and studied in a controlled environment to yield new insights into integrated physiology. In contrast, a large group of scientists believed that cells in culture were artifacts that could not truly represent processes going on in the whole animal.

A few key discoveries have been made in tissue culture, including the pathways of RNA metabolism in animals, the anomolous karyotypes in some human genetic diseases, and advances in animal virology. These discoveries did not depend on a perfect correlation between the culture model systems and the animals that were modelled. This illustrates an important principle: the more basic the question, the less perfect the model needs to be. The lack of confidence in tissue culture systems probably accounts for many lost opportunities during the early days of cell culture. This technique might have been used to discover many of the vitamins, amino acids, and hormones.

Today, the opportunities for cell culture to contribute to cellular immunology can be divided into two parts. The first is a mechanism of action of regulatory molecules. Cultures have a great advantage in that they are easy to manipulate and systems exist today. Therefore, certain questions can be addressed. For example, how does interleukin trigger the growth of the T cell, and how does an antigen bring about an increase of interleukin receptors? Questions of mechanism, since they are so basic, will not suffer from doubts about the artifactual nature of cell culture. The second area of opportunity is integrated physiology of immune systems. Here we will be nagged by doubts about how factors discovered in culture are relevant to whole animal immunology. Culture offers the best opportunity to discover these factors, but their relevance must be determined by experiments in animals. Some immune deficiency diseases may some day be traced to some disturbance in the production or response to these factors. Such a development, of course, would lend great credence to their validity. Another approach would involve animals in which the levels of these factors or receptors were experimentally manipulated. The final validation of these factors would be theoretical, and one could look forward to the day when a coherent picture of immunology can be constructed from these individual factors and their reactions.

REFERENCES

Acuto, O., R. E. Hussey, K. A. Fitzgerald, J. A. Protentis, S. C. Meuer, and E. L. Reinherz. 1983. The human T cell receptor: Appearance in ontogeny and biochemical relationship of alpha and beta subunits on IL-2 dependent clones and T cell tumors. Cell 34:717-726.

Claman, H. N., S. D. Miller, P. J. Conlon, and J. W. Moorhead. 1980. Control of experimental contact sensitivity. Adv. Immunol. 30:121-157.

Farrar, J. J., W. R. Benjamin, M. L. Hilfiker, M. Howard, W. L. Farrar, and J. Fuller-Farrar. 1982. The biochemistry, biology, and role of interleukin 2 in the induction of cytotoxic T cell and antibody-forming B cell responses. Immunol. Rev. 63:129-166.

Germain, R. N., and B. Benacerraf. 1980. Helper and suppressor T cell factors. Springer Sem. Immunopathol. 3:93-127.

Harrison, R. G. 1907. Observations on the living developing nerve fiber. Proc. Soc. Exp. Biol. Med. 4:140-143.

Levine, M. A., and H. N. Claman. 1970. Bone marrow and spleen: Dissociation of immunologic properties by cortisone. Science 167:1515-1516.

Metchnikoff, E. 1905. Immunity in infective diseases. F. G. Binnie, transl. Cambridge University Press, Cambridge, England. pp. 591.

Meuer, S., S. F. Schlossman, and E. L. Reinherz. 1982. Clonal analysis of human cytotoxic T lymphocytes: T 4 positive and T 8 positive effector T cells recognize products of different major histocompatibility complex regions. Proc. Natl. Acad. Sci. USA 79:4395-4399.

Meuer, S. C., R. E. Hussey, D. A. Cantrell, J. C. Hodgdon, S. F. Schlossman, K. A. Smith, and E. L. Reinherz. 1984. Triggering of the T 3-Ti antigen-receptor complex results in clonal T cell proliferation through an interleukin 2-dependent autocrine pathway. Proc. Natl. Acad. Sci. USA 81:1509-1513.

Mosier, D. E., and B. Subbarao. 1982. Thymus-independent antigens: Complexity of B-lymphocyte activation revealed. Immunol. Today 3:217-222.

Oppenheim, J. J., D. L. Rosenstreich, and M. Potter, eds. 1981. Cellular Functions in Immunity and Inflammation. Elsevier/North Holland, New York. 479 pp.

Phanuphak, P., J. W. Moorhead, and H. N. Claman. 1974a. Tolerance and contact sensitivity to DNFB in mice. I. *In vivo* detection by ear swelling and correlation with *in vitro* cell stimulation. J. Immunol. 112:115-123.

Phanuphak, P., J. W. Moorhead, and H. N. Claman. 1974b. Tolerance and contact senstivity to DNFB in mice. III. Transfer of tolerance with "suppressor T cells." J. Immunol. 113:1230-1236.

Rook, A. H., H. Masur, H. C. Lane, W. Frederick, T. Kasahara, A. M. Macher, J. Y. Djeu, J. F. Manischewitz, L. Jackson, A. S. Fauci, and G. V. Quinnan, Jr. 1983. Interleukin-2 enhances the depressed natural killer and cytomegalovirus-specific cytotoxic activities of lymphocytes from patients with the acquired immune deficiency syndrome. J. Clin. Invest. 72:398-403.

Ziegler, H. K., and E. R. Unanue. 1982. Decrease in macrophage antigen catabolism caused by ammonia and chloroquine is associated with inhibition of antigen presentation to T cells. Proc. Natl. Acad. Sci. USA 79:175-178.

Appendix C

Models for the Investigation of Learning

Workshop Program

May 30, 1984
Washington, D.C.
Michael V. L. Bennett
Organizer and Chairman

Bela Julesz AT&T Bell Laboratories Murray Hill, N.J.	Introductory Talk: What Can and Cannot Be Learned in Human Vision According to the Texton Theory
Thomas Carew Yale University New Haven, Conn.	Cellular Mechanisms of Associative Learning in *Aplysia*
William Quinn Princeton University Princeton, N.J.	Genetic Studies of Learning and Memory in *Drosophila*
Alan Gelperin Laboratories Murray Hill, N.J.	Cellular and System Level AT&T Bell Studies of Associative Learning in Protohumanoid Mulluscs
Howard Berg California Institute of Technology Pasadena, Calif.	Short-Term Memory in Bacterial Chemotaxis
Fernando Nottebohm Rockefeller University Millbrook, N.Y.	Song Learning in Birds as a Model for Biomedical Research
Leon Cooper Brown University Providence, R.I.	From Neuron Learning to Network Organization
Patrick Winston Massachusetts Institute of Technology Cambridge, Mass.	A Computational View of the Learning Process
Gary Lynch University of California, Irvine Irvine, Calif.	A Biochemical Hypothesis of Memory: A Description and Preliminary Test
Richard Thompson Stanford University Stanford, Calif.	Mammalian Models of Basic Associative Learning
Larry Squire University of California, San Diego La Jolla, Calif.	The Biology of Memory in Humans and Nonhuman Primates: A Neuropsychological Analysis

Workshop Summary

The study of learning demonstrates the value of comparative physiology, which loosely speaking might be considered the study of common aspects of organismic function in different species. The criteria for the presence of learning were derived from work on mammals, starting with Pavlov's dogs. Associative learning, although generally regarded as a higher nervous function, can be demonstrated in invertebrates as well as in lower vertebrates. The survival value of learning is not restricted to higher life forms, and in several learning tasks some lower vertebrates and invertebrates demonstrate considerable sophistication. The question then becomes whether the mechanisms are similar at the different phylogenetic levels. The answer derived from this workshop is: not necessarily.

In terms of the committee's charge, progress in learning about learning has required and will require experiments on a number of invertebrates and lower vertebrates. But at more integrative levels, there is often no alternative to studies in primates, including humans. According to Bela Julesz (AT&T Bell Laboratories), psychophysical phenomena in humans provide important guidance and questions for animal research. Studies of neurological patients shed light on normal function, which is further analyzed in animals; of course, these patients also provide the medical problems for which one seeks animal models.

Historically, the phenomenon of learning as a behavior was first carefully defined by experimental psychologists, who worked largely with mammals. As described by Thomas Carew (Yale University), application of these principles to the gastropod *Aplysia* (a genus of sea hare) led to the description of a simple circuitry that mediates behavior, satisfying all the criteria for associative learning (Carew *et al*., 1983). The achievement comes very close to identifying an engram -- the elusive memory trace. Now, the underlying mechanism -- use-dependent facilitation -- is being looked for in higher forms.

Genetic analysis suggests that biochemical pathways similar to those in *Aplysia* are involved in learning in the fruit fly *Drosophila* (Tempel *et al*., 1984). This was described in a presentation by William Quinn (Princeton University). By its nature, the analysis skips over the neural circuitry. However, the responses to drugs in the two organisms are sufficiently similar to suggest an underlying common mechanism. The biochemical analysis of behavioral mutants was inspired by the *Aplysia* work and extends its relevance.

Circuitry similar to the widely distributed reinforcing presynaptic input found in *Aplysia* has long been known to exist at several sites in the mammalian brain. The *Aplysia* model gives the somewhat peculiar morphology a new functional interpretation and is leading to experimentation that otherwise might not have been conceived for a much longer time. It appeared that the workshop participants who used mammals had been clearly stimulated by the invertebrate model users and were busily testing or at least thinking about the applicability of the investigation of invertebrates to their systems.

Conversely, mammalian studies have guided the investigation of other forms of learning in invertebrates. Alan Gelperin (AT&T Bell Laboratories) told the assemblage that learning of food preference exhibits some properties remarkably different from those of classical conditioning (Sahley *et al*., 1984). An aversive stimulus that causes vomiting can lead to the avoidance of a food that was consumed hours before the the stimulus was given. The temporal relationships are in sharp contrast to the interstimulus interval of seconds required for conditioning by nociceptive (i.e., painful) stimulation. The survival advantage of learning to recognize toxic foods is apparent. Hence, it is not surprising that a garden slug exhibits this kind of learning, which was first observed in rats by Garcia *et al*. (1974).

Another phenomenon in food preference learning is called block of conditioning. If a food perceived by the animal as good is mixed with a novel food and the aversive stimulus is given, only the novel food is avoided in subsequent tests. If the good food alone is paired with the aversive stimulus, it is subsequently avoided. The novel food protects the familiar good food against the learned association with the aversive stimulus. Again, the survival advantage to an animal on a mixed and variable diet is obvious. Although use of the model system was inspired by prior study in a higher form, analysis at the cellular level is likely to be accomplished sooner in the lower organism because of the greater simplicity of its brain. It will then be necessary to validate the model's applicability to mammalian forms. Thus, as demonstrated by the examples given here, there are common elements to learning not only among lower organisms themselves but also between higher and lower systems.

It may seem absurd to talk about learning in unicellular organisms, especially since learning in higher organisms is described as requiring cellular interactions. Nonetheless, as discussed by Howard Berg (California Institute of Technology), very simple learning can be observed in bacteria -- not associative learning but, rather, changes in responsiveness resembling sensory adaptation (Hazelbauer and Harayama, 1983). A bacterium detects the concentrations of many different substances in its environment and compares them to the concentrations present a few seconds previously. If the concentration of a food substance is increasing, the bacterium tends to continue swimming in the same direction. If it is not, the bacterium tends to randomly change its orientation.

There is some understanding of the short-term memory involved in this behavior at the molecular level, but this knowledge is not yet complete. The chemical receptor, a transmembrane macromolecule, is slowly methylated on the cytoplasmic side when occupied by the attractant molecule on the external side and is rapidly demethylated when not occupied by the attractant. Comparing receptor occupation (degree of methylation) in the more distant past with receptor occupation in the present and more recent past "tells" the bacterium whether it is going toward or away from the food source. There is as yet no synaptic process that is known to operate by similar mechanisms. However, sensory adaptation, or habituation, may (or does) involve analogous covalent modification of membrane macromolecules by cytoplasmic protein kinases (Kandel and Schwartz, 1982). Furthermore, the transmembrane signalling by membrane molecules in bacteria may provide important insights into operation of analogous (or homologous) molecules in excitable membranes and other cells.

The study of lower organisms can bring to light neural mechanisms usually assumed not to occur in higher forms. As discussed by Fernando Nottebohm (Rockefeller University), male canaries and a number of other song birds forget their songs in the summer at the end of the mating season and relearn them every spring (Goldman and Nottebohm, 1983). This sequence involves first regression and then regrowth in the controlling region of the brain. The surprising finding is that, during the regrowth phase, new neurons are formed from the ependyma -- the epithelial membrane lining the ventricles of the brain and the canal of the spinal cord. A variety of evidence indicates that such continued formation of neurons in the adult does not occur in mammals. An interesting implication is that specific behavior requires its own neural circuitry occupying its own space. Forgetting involves loss of neurons, and new learning involves growth of new neurons and connections. It is commonly believed that our memory banks do not fill up, although slower learning, at least of detail, is often observed in older people.

Of major concern in neurology are the irreversible effects of lesions in the central nervous system (CNS). The birds provide the most phylogenetically advanced example of replacement of CNS neurons. To control the process, or to learn how to induce the replacement process in humans, conceivably would lead to useful treatment of spinal cord injury, stroke, and other traumas. The possibility seems remote at this time, and bird brains may be too complex to permit experimental manipulation of the cellular controls that allow neuronal replacement. Still, the possibility is exciting.

An important point with respect to the study of nonmammalian organisms as models is that the resultant observations may be unexpected. For example, birds have attracted watchers and ethologists for a long time, but it is very unlikely that they were initially considered as models for CNS regeneration. The extent to which birds have been studied as models of learning in higher vertebrates was not considered in the workshop.

The committee's charge includes consideration of mathematical models. One can argue that understanding a process or phenomenon requires a mathematical or symbolic description. In this sense, all quantitative understanding involves mathematical models. It was agreed by the two mathematical presenters that models as hypotheses should stay as close as possible to experimental data -- both anatomical and physiological. Nevertheless, Leon Cooper (Brown University) described certain kinds of memory that could be exhibited by a hypothetical neuronal circuit, and a number of experiments in animals supported its possible validity (Cooper, in press).

When it comes to more complex functions, such as intelligence, Patrick Winston (Massachusetts Institute of Technology) related that attempts to make a smart machine (or expert computer program) soon lead to a greater understanding of the complexity of ordinary thought processes in humans (Winston, 1984). Humans and animals do a great deal of parallel processing, as in vision and in other perceptual processes. Computers at this time are largely serial, and although their single steps are much faster, animals solve perceptual problems more rapidly. Although there is no reason to doubt that a thinking machine will be built, that outcome does not appear imminent. In terms of analysis of neural function, computers are likely to be more useful when the investigators try to understand how the animal performed a problem-solving task rather than when they ask how they would solve it themselves if they had the same problem as the animal. The latter approach may well have value for simple processes, but is less likely to be useful when complex behavior is involved.

Three of the speakers dealt with mammalian systems. By using brain slices, as described by Gary Lynch (University of California, Irvine), the brain from the intact animal can be rapidly removed and divided into responding, but clearly not sentient, fragments. Brain slices provide several important advantages for cellular analysis. The preparation is not affected by respiration or by circulatory pulsation, its chemical environment can be controlled, and its chemistry can be studied. Lynch and Baudry (1984) have used brain slices to investigate long-term potentiation, i.e., increased transmission at certain synapses as a result of prior activation. The potentiation lasts for weeks and has been proposed as a substrate for memory. Because of experimental simplicity in the slice, Lynch was able to suggest the mechanism of the potentiation. Then with knowledge from the *in vitro* preparation, he went back to the intact organism and obtained pharmacological evidence for the same physiology. He could also show that a pharmacological block of long-term synaptic potentiation affected certain classes of learning, leaving others unaffected. In this system *in vitro* studies were required to understand the mechanism, but a return to the intact organism was necessary to give it physiological meaning. The mechanism appears to be absent in lower vertebrates; if so, the study of mammals is essential to its description.

Simple systems are found not only in lower organisms but also sometimes in mammals. These were described by Richard Thompson (Stanford University). The eye blink can be conditioned in rabbits, and the neuronal site for this reflex appears to be located within a 1-mm^3 region in the interpositus nucleus of the cerebellum. The number of neurons involved is still large, but the circuitry is certainly simplified, compared to the cerebral cortex. *In situ* or *in vitro* analysis at the cellular level may be possible. Here the invertebrate data provide guidance for the initial studies to be performed in mammals (Thompson, 1983).

An important point emphasized by Larry Squire (University of California, San Diego) is the necessity of studying higher order learning in higher organisms. One can now observe in primates effects of lesions that correspond to amnesia, and the pathways of different kinds of learning are being described. Lesions seen by the neurologist provide one kind of data, and the availability of computer-assisted tomography (CAT) scan and nuclear-magnetic resonance (NMR) imaging will make interpretation and diagnosis of these cases much more rapid and their study, therefore, more attractive. But at present there is no substitute for the accurately placed surgical lesion in the brain of an infrahuman primate in elucidating pathways and effects on behavior. Moreover, data collected from brain slices must be verified with data taken from conscious and behaving animals (Squire, 1982).

The participants in this workshop thus pointed to a number of specific applications of models in studies of learning:

- Some mechanisms of learning may be revealed more readily in lower organisms than in higher forms because of the structural simplicity of the former.

- However, the questions asked of these organisms are often derived from previous study of higher organisms. Study of the same processes in different animals at different phylogenetic levels is mutually facilitative.

- Study of lower organisms to illuminate one problem (or for its own sake) may lead to quite unexpected observations that are relevant to other problems.

- Mathematical models or treatments of physiological processes cannot replace physiological study but are natural complements thereof and may provide important guidance.

• *In vitro* preparations from mammalian tissues are, like lower organisms, useful and can provide insights into mechanisms. However, relevance of the resultant data to behavior must be tested in the intact organism.

• When it comes to primate and human behavior, one learns a great deal from the study of primates and humans. Indeed, one must study them, however much one learns about mechanisms of learning in simpler organisms.

REFERENCES

Carew, T. J., R. D. Hawkins, and E. R. Kandel. 1983. Differential classical conditioning of a defensive withdrawal reflex in *Aplysia californica*. Science 219:397-400.

Cooper, L. N. In press. Neuron learning to network organization. Proceedings of the Maxwell Symposium. Elsevier/North Holland, Amsterdam, The Netherlands.

Garcia, J., W. G. Hankins, and K. W. Rusinak. 1974. Behavioral regulation of the milieu interne in man and rat. Science 185:824-831.

Goldman, S. A., and F. Nottebohm. 1983. Neuronal production, migration and differentiation in a vocal control nucleus of the adult female canary brain. Proc. Natl. Acad. Sci. USA 80:2390-2394.

Hazelbauer, G. L., and S. Harayama. 1983. Sensory transduction in bacterial chemotaxis. Int. Rev. Cytol. 81:33-70.

Kandel, E. R., and J. H. Schwartz. 1982. Molecular biology of learning: Modulation of transmitter release. Science 218:433-443.

Lynch, G., and M. Baudry. 1984. The biochemistry of memory: A new and specific hypothesis. Science 224:1057-1063

Sahley, C., J. W. Rudy, and A. Gelperin. 1984. Associative learning in a mollusk: A comparative analysis. Pp. 243-258 in D. C. Alkon and J. Farley, eds. Primary Neural Substrates of Learning and Behavorial Change. Cambridge University Press, New York.

Squire, L. R. 1982. The neuropsychology of human memory. Annual Rev. Neurosci. 5:241-273.

Tempel, B. L., M. S. Livingstone, and W. G. Quinn. 1984. Mutations in the DOPH-decarboxylase gene block learning in Drosophila. Proc. Natl. Acad. Sci. USA 81:3577-3581.

Thompson, R. F. 1983. Neuronal substrates of simple associative learning: Classical conditioning. Trends. Neurosci. 6:270-275.

Winston, P. H. 1984. Artificial Intelligence. Addison-Wesley. Reading, Massachusetts. 527 pp.

Appendix D

Nonmammalian Models for the Study of Biological Regulation

Workshop Program

June 15, 1984
Washington, D.C.
Lawrence I. Gilbert
Organizer and Chairman

Jesse Roth National Institutes of Health Bethesda, Md.	Introductory Talk: Materials and Microbes That Resemble Vertebrate Hormones and Neuropeptides: Evolutionary Origins of Intercellular Communication
Jay Dunlap Dartmouth College Hanover, N.H.	Biological Clocks in *Neurospora*, *Drosophila*, and *Gonyaulax*
Phillip Sokolove University of Maryland, Baltimore County Catonsville, Md.	Molluscan Hormones and Photoperiodism
James Truman University of Washington Seattle, Wash.	Environmental Events and Hormones in Invertebrates
Berta Scharrer Albert Einstein College of Medicine Bronx, N.Y.	Evolutionary History of Peptidergic Neurons
John Hildebrand Columbia University New York, N.Y.	Development of the Central Nervous System in a Metamorphosing Insect -- Influences of Peripheral Inputs on Central Neural Development
Thomas Miller University of California, Riverside Riverside, Calif.	Neuropharmacology in Insects
Bruce Hammock University of California, Davis Davis, Calif.	Insects as a Model System in Endocrinology and Toxicology
Earl Frieden Florida State University Tallahassee, Fla.	Amphibian Metamorphosis: A Model for Hormonal Regulation and Controlled Differentation

Workshop Summary

All eucaryotic cells and organisms have the capacity for endogenous control and organization over time. As described by Jay Dunlap (Dartmouth College), the cellular machinery that generates this ability is what is collectively known as the biological clock. The importance of a detailed understanding of the circadian biological clock rests on the apparent ubiquity of its influence on cellular and organismal processes. Phylogenetically, this ranges from control of cell division and enzyme activities in unicells to plant photoperiodism, avian and insect celestial navigation, and a multiplicity of human systems, including endocrine function, work-rest cycles and sleep, and drug tolerances and effectiveness (see Aschoff, 1981).

In the last 20 years significant progress has been made in identifying the locus of circadian pacemakers. They are usually found in the head, and the tissues implicated in pacemaker function are often neuronal (Jacklet, 1981). Since cells are known to communicate, the feedback cycle comprising the pacemaker could lie at the level of either intercellular communication or intracellular regulation. In the _Aplysia_ eye, ionizing radiation has been used to calculate an approximate upper limit (about 10^8 cubic angstroms) for the size of the target that must be hit to inactivate the clock (Woolum and Strumwasser, 1980). This corresponds to a sphere roughly 0.06 μm in diameter, which is considerably smaller than a cell. In addition, efforts to find light-entrainable circadian oscillations in dissociated cell culture have been successful both with chicken pinealocytes and _Drosophila_ embryonic cells. Finally, several drugs and metabolites have been shown to affect a neuronal clock (the _Aplysia_ eye) at concentrations that do not significantly affect intercellular communication (Jacklet, 1981). These findings clearly suggest that metazoan model systems provide fertile ground for the study of the clock at the level of intracellular regulation and that, as a first approximation, unicellular and microbial systems may be good models even for vertebrate clocks.

In studies of the biological clock mechanism at the cellular level, a major thrust has been to establish the identity of one of its components -- a gear or cog in the clock mechanism. This has not yet been possible in any system, but in general one of three orthogonal approaches has been taken (Feldman, 1982). In one, the information flow is followed either to the clock mechanism (by studying entrainment) or from it (by studying controlled processes, i.e., the hands of the clock). This approach has been used most extensively in examining the clock's bioluminescence control of the unicellular algae _Gonyaulax_ (Dunlap and Hastings, 1981).

Light emission is the result of a single enzyme-substrate reaction, both components of which are controlled by the clock. This reaction provides a noninvasive probe into clock control of the cell's internal biochemistry.

Purification of the enzyme luciferase has enabled investigators to produce specific antisera and prove that luciferase activity per unit luciferase protein is the same at the high and low points of the oscillation. This strongly suggests that the actual amount of luciferase must be changing throughout the day and that the rhythm is thus controlled at the level of synthesis and degradation (Johnson *et al.*, 1984). Experiments are now in progress to clone the luciferase gene for use in studying the level at which synthesis or degradation is controlled.

As reviewed by Dunlap, work on a variety of organisms, including the unicellular algae *Gonyaulax* and *Acetabularia*, the mollusc *Aplysia*, and the mold *Neurospora*, has shown that even treatment of very short duration with inhibitors of protein synthesis such as cycloheximide can permanently reset the clock. The use of mutant strains of *Neurospora* whose ribosomes are resistant to cycloheximide has proven that the resetting effect is actually mediated by the inhibition of protein synthesis, as distinct from being a side effect of the drug. There are other mutant strains whose ribosomes have a normal sensitivity to cycloheximide but whose clock is partially resistant to resetting. Thus, at least for *Neurospora*, proteins that turn over rapidly probably must be resynthesized daily to maintain normal clock function, but that daily resynthesis *per se* cannot be a step in the feedback loop (reviewed by Feldman and Dunlap, 1983).

In the third approach, the genetic basis of clock characteristics has been examined. The isolation and characterization in several systems of single gene mutants affecting such basic clock characteristics as period length, light entrainability, and temperature compensation has left no doubt that clocks can be subjected to classical genetic analysis (Feldman, 1982). A number of surprises emerged from this approach, including the following:

- a high proportion of single gene mutants frequently map to the same genetic loci (*per* in *Drosophila*; *frq* in *Neurospora*);

- the effects of unlinked mutations in the same organism are roughly additive;

- mutant phenotypes are frequently codominant with, instead of recessive to, the wild phenotype;

- single loci can be mutated to give long, short, and arrhythmic clocks; and

- single gene mutants often affect more than one basic clock property (e.g., period length, temperature compensation, or light entrainability).

These data suggest an interrelatedness among clock properties at the molecular level that would not have been predicted *a priori*. A most unexpected finding is that the courtship song rhythm of *Drosophila* has at least one of the same genetic determinants as the circadian clock (Kyriacou and Hall, 1980). Through mosaic analysis, the thorax has been pinpointed as the source of control of the courtship song. Apparently, the same gene(s) or gene product(s) are being used in separate tissues to time different processes with grossly different time constants. Efforts now in progress to clone *frq* and *per* genes (Bargiello and Young, 1984; Reddy *et al*., 1984) will greatly facilitate the examination of the products and regulation of these genes.

Another example of an environmentally controlled event is reproduction. Recent work suggests important parallels between molluscan and mammalian systems that should encourage the use of certain snails and slugs as models for photoperiodic regulation of reproduction. Phillip Sokolove (University of Maryland, Baltimore County) discussed two species of pulmonates -- the terrestrial slug *Limax maximus* and the pond snail *Lymnaea stagnalis*. These two organisms probably represent the clearest opportunities for the development of models related to photoperiodism and reproductive endocrinology. Other molluscs, such as *Aplysia* and the Japanese snail *Euhadra peliomphala*, also offer exciting possibilities as models for peptide and steroid reproductive hormone action, but photoperiodic regulation of reproduction in these species has not yet been demonstrated.

The hypothalamo-hypophyseal-gonad axis is the generally accepted model for the control of vertebrate reproductive systems. Hypothalamic releasing factors promote the release of follicle-stimulating hormones and luteinizing hormones, and these peptides regulate gametogenesis and gonadal steroid hormone production. Gonadal hormones, in turn, control the development of secondary sexual characteristics and play an important intragonadal role in stimulating gamete development. The effect of day length on this system is not especially notable in humans, but in many other mammalian species changing photoperiod can dramatically affect the reproductive state by indirectly altering the production of pituitary hormones.

In *Limax maximus*, the brain-gonad axis has been shown to regulate gametogenesis and reproductive tract development. Research to date indicates that neurosecretory cells in the cerebral ganglia of the slug produce a protein maturation hormone, or male gonadotropic factor (MGF), that stimulates spermatogonial proliferation and promotes secretion of one or more gonadal hormones (McCrone and Sokolove 1979; McCrone *et al*., 1981; Sokolove *et al*., 1983). Gonad hormone induces growth and development of accessory sex organs (ASO), such as the penis apparatus and the albumin gland. It may also regulate spermiogenesis intragonadally. When slugs are exposed to long daylight cycles, secretion of the brain hormone appears to be permanently triggered, which initiates the sequence of events associated with reproductive maturation (McCrone and Sokolove, 1979; Sokolove *et al*., 1984).

As a potential model for vertebrate systems, *Limax* has both advantages and disadvantages. Our knowledge of its endocrine system is still very incomplete. Neither brain nor gonadal hormones have yet been isolated and identified, and evidence is lacking for the functional equivalent of an anterior pituitary. Also, like other pulmonates, *Limax* is a hermaphrodite, and the regulation of reproductive events may therefore be both very different and more complex than in species with separate sexes. Nevertheless, the robustness of the slug brain's secretory response to long daylight hours suggests that MGF cells can be used to investigate photoperiod-induced neuroendocrine events at the cellular and perhaps even molecular level. Furthermore, since slug albumin gland explants seem to develop normally when cocultured with gonads, one may also be able to study *in vitro* hormone-induced cellular proliferation and morphogenesis in intact glandular tissue.

Emerging data are providing more details about the endocrine and neuroendocrine centers that regulate reproductive development and activity in the pond snail. Specific cell types have been identified functionally as well as histochemically, and rapid strides are being made to isolate active secretory products. The neuroendocrine caudodorsal cells (CDCs), in particular, have been studied intensively. Like the bag cells in *Aplysia*, the CDC in *Lymnaea* produce a peptide hormone that regulates ovulation and ovipository behavior (Geraerts and Bohlken, 1976).

Egg producton in *Lymnaea* depends on a number of environmental factors, including temperature, food availability, and water quality, but the dominant factor is photoperiod (Joosse, 1984), especially when other conditions are adverse. Long days increase both oviposition frequency and the number of eggs per clutch. At the cellular level, long days also lower the CDC threshold for electrical stimulation during 30 to 60 minutes after the discharge that accompanies hormone release (de Vlieger *et al.*, 1980; Kits, 1980). The CDCs appear to provide an ideal model system for the study of environmental regulation of neurosecretory processes and for investigating the physiological and behavioral actions of reproductive neuropeptides.

As a whole, the *Lymnaea* reproductive endocrine system exhibits both similarities and differences in comparison with vertebrate systems. Virtually all aspects of female development, including vitellogenesis, cellular differentiation, and growth and synthetic activity of female ASO (Geraerts and Joosse, 1975), are regulated by the hormone of the endocrine dorsal bodies, which therefore might be considered the functional equivalent of the vertebrate anterior pituitary. On the other hand, the gonad is not implicated in any reproductive endocrine role whatever.

Although not photoperiodic, the slug *Ariolimax californicus* and the snail *Euhadra* probably deserve special mention, because androgens promote spermatogenesis and male phase development in these species (Takeda, 1983; Gottfried and Dorfman, 1970), whereas estrogens stimulate oogenesis and female development (Takeda, 1983). These pulmonates may be useful in studies of the mechanisms of action of vertebrate steroid hormones that appear to retain their familiar functions in such molluscs.

Insects are also excellent models for studying the transduction of environmental events into hormonal signals. As described by James Truman (University of Washington, Seattle), their endocrine system is composed of glands that secrete steroid and sesquiterpenoid hormones and of neurosecretory cells that release a variety of peptide hormones. Environmental factors regulating endocrine activity act primarily through the neurosecretory pathways. Some act through reflexive control systems, in which a particular stimulus directly triggers an endocrine response e.g., the triggering of diuretic hormone release by the ingestion of a blood meal by *Rhodnius*, whereas others act through endogenous control systems, such as those associated with circadian rhythms or with photoperiodism. Some of the major problems in the regulation of endocrine activity relate to the latter types of control systems. Insects with their rather simple and accessible nervous systems are useful animals for studying this type of action.

The first model relates to the photoperiodic regulation of endocrine function -- an issue that is poorly understood at the mechanistic level. In the tobacco hornworm (*Manduca sexta*), the prothoracicotropic hormone (PTTH) is released from certain brain cells and acts on the prothoracic glands to cause the secretion of ecdysone, which in turn brings about molting (Gilbert *et al.*, 1980). At the pupal stage, the moth either immediately begins to develop as an adult or enters a state of developmental arrest (diapause). This decision is controlled primarily at the level of PTTH release (Agui *et al.*, 1979; Gilbert *et al.*, 1981) and is determined by the photoperiod experienced by the insect earlier, during larval life. Brains can be taken from larvae, placed in short-term culture, and "programmed" either for diapause or development, depending on the photoperiod experienced while in culture (Bowen *et al.*, 1984). This, therefore, provides a unique model system in which photoperiodic decisions can be studied in the context of identified neurosecretory cells under well-controlled *in vitro* conditions.

The release of many hormones is associated with circadian "clocks," which modulate the daily secretion of hormones or determine which hormone is released, thereby confining secretion to a certain time on a particular day (Porter and Collins, 1984). Environmental factors, such as light and temperature, work through these internal clocks to synchronize endocrine activity with the 24-hour day. These circadian control systems are complex, consisting of a number of oscillators (associated with both neural and endocrine components) assembled in a hierarchical arrangement that is not well understood. The coupled oscillator models for the circadian system originally were derived from studies of the eclosion rhythm of *Drosophila* by Pittendrigh *et al*. (1984). Subsequent studies on moths showed that this particular rhythm is based on the gated release of a peptide hormone (the eclosion hormone) from specific cells in the brain of the moth (Truman, 1984). Recent studies on the hawkmoth *Manduca sexta* described by Truman indicate that eclosion is regulated by the interaction of two centers: a light-sensitive clock located in the brain of the moth and a temperature-sensitive clock located outside the brain that communicates with the brain system through daily modulations in the ecdysteroid titer. The site of this second clock may be the prothoracic glands themselves, but this has not yet been determined. The system can be used to study the manner in which two components of a circadian system interact to determine the final expression of the rhythm, since the final output of the system is through an identified neurosecretory cell whose inputs and functions can be studied electrophysiologically in dissected preparations.

One of the most difficult aspects of hormone action to study is the manner by which hormones alter the functioning of the central nervous system. The best vertebrate model is the rat, especially for the hormonal regulation of sexual behavior. However, the central nervous system of the rat and other mammals is very complex and thus difficult to study, a problem that is somewhat overcome by the simpler nervous systems and functionally defined neurons of invertebrates. The effect of hormones on behavior is particularly striking in arthropods, in which hormones used as neural modulators mediate the programming of a variety of behaviors with a relatively modest amount of neuronal circuitry. This principle is nicely demonstrated by the effects of various modulators on the stomatogastric ganglion of lobsters (Kravitz *et al.*, 1983). This ganglion is composed of only 30 neurons (all but one being motoneurons) that generate various stereotyped rhythmic patterns. Recent studies have shown that these chemicals strengthen some and weaken other synapses in the network, thereby deriving a functional subset of neurons from the pool of available cells. The same network can thus generate different complex rhythmic patterns, depending on the modulator that is present.

In *Manduca*, the eclosion hormone acts on the central nervous system, triggering the motor pattern seen at ecdysis and activating a number of neural pathways that are necessary for the behavior of the postecdysis insect (Truman *et al.*, 1981). A number of these hormone-sensitive circuits have been studied, and possible sites of peptide action have been identified. They provide models for studying some of the cellular actions that may accompany changes in synaptic function. At present, this model suffers from the fact that eclosion hormone has not been sequenced and some of the ligands that would be useful in probing the action are not yet available.

Another aspect of the eclosion hormone model system is that the central nervous system responds to the peptide only after it has been exposed to ecdysteroids (insect molting hormones) (Schwartz and Truman, 1983). An indication of the biochemical changes underlying the onset of sensitivity is seen in the changes in target cell cyclic GMP (guanosine cyclic 3',5'-hydrogen phosphate), a putative second messenger for eclosion hormone. During the acquisition of sensitivity, the target cells progress through a stage during which they will not respond physiologically to the eclosion hormone but will respond by showing their characteristic increase in intracellular cyclic GMP. These findings suggest that regulation of cellular responsiveness may lie at other levels besides the hormone receptor. Thus, this model be useful in the study of regulation at this level.

The invertebrate nervous system has also been a valuable model in elucidating basic phenomena that are characteristic of the mammalian nervous system (see Chapter 4). The use of invertebrate models to study the nature and diverse operational modes of peptidergic neurons certainly has been rewarding. These nerve cells, which make use of peptides (Gainer, 1977) for signalling to other cells in various ways (e.g., by neurohormonal, transmitterlike, and modulatory signals), have a long evolutionary history antedating that of conventional, synaptically transmitting neurons (Haynes, 1980). Therefore, the examination of invertebrate systems as far down as the most primitive multicellular animals, the coelenterates, provides us with factual and conceptual insights that could not be gained from the study of vertebrates alone.

Moreover, the remarkable parallel between the structure and function of peptidergic neurons and those of neuron systems throughout the animal kingdom allow us to use invertebrate models effectively in the search for basic principles. Berta Scharrer (Albert Einstein College of Medicine) pointed out that such parallels have been recognized from the very beginning of this research more than 50 years ago. The discovery of neurohormone-producing "neurosecretory neurons" in the vertebrate hypothalamus was soon followed by discovery of analogous neurons in annelids, molluscs, and arthropods (Scharrer, 1976, 1978a,b, 1981).

Insects were subsequently found to possess a neuroendocrine organ complex (the pars intercerebralis-cardiacum-allatum system). It is comparable to the hypothalamic-hypophyseal system of vertebrates, in which neurosecretory neurons provide the all-important link between the nervous and endocrine systems (Gilbert _et al._, 1980). Electron microscopy revealed the presence in both of electron-dense secretory granules, their formation in the Golgi apparatus, their storage in neurohemal organs (posterior pituitary, corpus cardiacum), and their release by exocytosis (Scharrer, 1981).

Immunocytochemical tests have indicated that a large number of vertebrate-type neuropeptides occur in invertebrates and vice versa (Duve and Thorpe, 1984; Kramer, in press). Two of the invertebrate neuropeptides thus far sequenced -- from a coelenterate (_Hydra_) and from a mollusc (_Mytilus_) -- are identical to their counterparts in mammals. Finally, neuropeptide receptors closely related to those of mammals have been detected in effector organs (e.g., ganglia and digestive tract) of two invertebrates -- the mollusc _Mytilus_ and the insect _Leucophaea_ (Stefano and Scharrer, 1981; Stefano _et al._, 1982).

Furthermore, the nervous system of invertebrate animals such as molluscs, worms, and insects have provided fertile ground for experimental studies of the functional organization, physiology, development, and behavioral roles of sensory, integrative, and motor centers and the elements they comprise. As we learn more and more about the cellular and molecular biology of nervous systems, it becomes clearer that basic structures, molecular components, and cellular mechanisms underlying the functions of the nervous system are essentially universal. Evolution seems to have varied the strategies or rules for assembling neural networks, but the elements of cellular circuitry and the molecular mechanisms responsible for electrical and chemical signalling in the nervous system have been conserved to a remarkable degree.

Perhaps representative of contemporary invertebrate neurobiology is the study by John Hildebrand (Columbia University) of a behaviorally important and experimentally tractable subsystem in the nervous system of an insect. His approaches have ranged from comparative physiology to neuroethology and from synaptic neurochemistry to developmental biology. Such invertebrate model systems permit the detailed anatomical, physiological, biochemical, developmental, and functional description that must underlie mechanistic inquiry as well as the controlled search for fundamental mechanisms and principles that build upon the essential description. One great advantage of invertebrate systems is their ready amenability to such multidisciplinary study.

Hildebrand's research is focused on the olfactory system of the hawkmoth *Manduca sexta*. He has explored in detail the cellular organization, cellular and synaptic physiology, synaptic neurochemistry, postembryonic development, and behavioral roles of this important sensory system. The rigorously applied cross-disciplinary experimental strategy used in these studies exploits technical advances in many fields. Studies of the development of the antennal olfactory pathway in *Manduca* have resulted in provocative findings concerning the role of cell interaction in developing neural pathways and the influence of sensory inputs on the development of target centers in the central nervous system (Hildebrand, in press; Hildebrand *et al*., 1979, 1982; Matsumoto and Hildebrand, 1981; Maxwell and Hildebrand, 1981; Schneiderman *et al*., 1982; Sanes *et al*., 1976; Sanes and Hildebrand, 1976).

Studies of sexually dimorphic elements in the olfactory center of the *Manduca*'s brain have revealed that male-specific, morphologically distinct types of neurons develop only in brains contacted by male-specific olfactory nerve fibers growing inward from the antennae. Moreover, a prominent synaptic neuropil structure (the macroglomerular complex) likewise develops only in antennal lobes receiving male olfactory nerve fibers. This primary olfactory center of the male brain (the antennal lobe) is the neuropil region in which male-specific sensory fibers form synapses with the dendrites of male-specific central neurons. Most remarkably, these male-specific features of the central nervous system develop in a genetically female antennal lobe if it is innervated by primary sensory fibers from a grafted male antenna. From this line of study, Hildebrand and his colleagues have concluded that cell to cell contact between male olfactory axons and their target neurons controls neuronal development. Their research is now seeking the molecular mechanism of this contact-mediated developmental regulation.

The basic principles of cell contact-mediated developmental regulation can be elucidated with this system. Just as fundamental insights about the hormone control of development have come from studies of metamorphosing insects, so an understanding of the mechanisms through which cells influence each other's development by contact-mediated processes may be attained with this model.

According to Thomas Miller (University of California, Riverside) among the many advantages of studying neuropharmacology in insects are their well-understood genetics, temperature adaptation, proliferation, lack of immunorejection, and the ability to withstand dissection, ablation, or axotony. Because insects do not always regulate temperature, they adapt to extreme temperature fluctuations -- a feature that provides some advantages to the experimenter. Furthermore, their immune system can be manipulated experimentally and segments of the insect can be studied individually because whole behaviors are observed in isolated parts of the body (Evered *et al*., 1982).

The neurochemical organization of insects seems to be the inverse of that in mammals. Their peripheral neuromusculature is a neuropeptide-aminergic system, whereas the system in vertebrates is central. Therefore, insects provide preparations for studying aminergic neurotransmission at single, identified cells. In studies of the vertebrate central nervous system, investigators must rely on averaging techniques to secure neuropharmacological information. Results of recent studies on biogenic amines are having a far-reaching impact on neuroethology, neuropharmacology, and neurotoxicology and are being followed by groups interested in both basic and applied research (Bullock, 1981; Strausfield and Adams, 1983).

The concept of insects as comparative and correlative, rather than predictive, models in endocrinology and toxicology is illustrated by the research of Bruce Hammock (University of California, Davis), who has control of larval-pupal transformation in lepidopteran insects by variations in rates of juvenile hormone catabolism (Hammock, in press; Ota and Hammock, 1980). Metamorphosis in higher insects has long been used as a model in developmental and regulatory biology (Gilbert *et al*., 1980). Basic research in insect endocrinology has helped to solve problems in primates in such diverse fields as neurotoxicity and carcinogenicity.

Depending on the species, insect juvenile hormone is catabolized by one of two hydrolytic enzymes: a highly selective esterase, which hydrates a α, β-conjugated methyl ester, or epoxide hydrolases, which hydrate the trisubstituted epoxide. In lepidopterous larvae, this juvenile hormone esterase appears twice during development: once apparently under neuroendocrine control and once under epithelial endocrine control. Hammock is testing a hypothesis that the first peak of the esterase is necessary not only for a reduction in biosynthesis but also for the reduction in juvenile hormone titer needed to initiate the pupation process. Of the several approaches that lend support to the hypothesis, the most straightforward one involves the synthesis of a new class of potent esterase inhibitors, which apparently act by mimicking the transition state of the enzyme. These ß-thiol trifluoromethylketones react stoichiometrically with the juvenile hormone esterase and allow one to titer it *in situ*. They also delay pupation *in vivo* (as expected of a juvenile hormone application) and delay the decrease of the hormone in the insect, as determined by gas-liquid chromatography-mass spectrometry. Affinity columns based on these compounds have led to a 1,400-fold purification of the esterase in one step.

A more direct application is used to study delayed neuropathy in humans. This syndrome has permanently afflicted entire villages, yet the chicken is the only laboratory animal that is a useful predictive model. Delayed neuropathy has been correlated with the permanent inhibition of the neurotoxic esterase (Abdel-aal *et al*., 1984). Interestingly, the same trifluoroketone is the most potent inhibitor of both the juvenile hormone esterase and the neurotoxic esterase in the human brain. Since this trifluoroketone is a slowly reversible inhibitor and is approximately 1,000 times as active as the insecticide mipafox, a dangerous delayed neurotoxic agent, it may find use as an antidote or in the purification of the neurotic esterase by affinity chromatography. Certainly a soluble blood enzyme in insects is not a useful predictive model for a membrane-bound esterase in the human brain; however, as a comparative model, the system has been very valuable.

Other insects use epoxide hydrolases to degrade the juvenile hormone, and in the fruit fly *Drosophila melanogaster*, epoxide hydrolases superficially resemble those of some vertebrates. Epoxide hydrolases of vertebrates have been studied intensively because of their ability to detoxify carcinogens and mutagens. However, during metabolism studies of insect hormones in vertebrates, a completely new class of these hydrolases was found. It had been thought that the epoxide hydrolases were entirely membrane-bound in mammals, but investigators found a soluble enzyme that had a wider substrate preference and a higher activity than the microsomal enzyme. Among the best substrates for the enzyme are epoxides of arachidonic acids, which may be chemical mediators, and epoxides of squalene and lanosterol, which are produced during steroid biosynthesis. The cytosolic epoxide hydrolase can reduce the mutagenicity of some epoxides in the Ames *Salmonella* assay as well.

The value of comparative work does not end here, because Hammock has used knowledge of the chemistry of insect hormones to prepare an affinity column for the cytosolic epoxide hydrolases from vertebrates. This column has led to the one-step purification of these hydrolases from several organisms, including rhesus monkeys and humans.

Hammock stated that there are good predictive insect models for endocrinological and toxicological events in higher vertebrates (Schuler *et al*., 1982; Walton, 1983). As we learn more about the mechanisms of hormone and toxin action, more such predictive models will be found. However, the major value of insect model systems is, and will be, their utility as comparative and correlative models.

Many nonmammalian vertebrate systems have been exploited as mammal surrogates, especially in studies of comparative endocrinology and developmental biology. Among the most important are the amphibians, which were discussed by Earl Frieden (Florida State University).

Studies of amphibian metamorphosis can shed light on many important metabolic and endocrinologic phenomena observed in higher vertebrates because of some basic similarities between the two (Gilbert and Frieden, 1981). For example, the endocrine control of the thyroid gland in amphibians from the order Salientia has the same fundamental components as in mammals. In this order, which includes tailless adults such as frogs and toads, a variety of external or internal stimuli can initiate neuroendocrine control in the hypothalamus, causing the secretion of thyrotropin-releasing hormone (TRH), which in turn stimulates the anterior pituitary to release thyroid-stimulating hormones (TSH). TSH regulates the output of T3 and T4 by the thyroid gland. The ultimate response of the peripheral tissue depends on its receiving thyroid hormone from the binding proteins in the serum and its capacity to respond (Galton, 1984; Yoshizato and Frieden, 1975). This culminates in a sharp increase in plasma T3 and T4, which occurs concomitantly with the transition of the tadpole to the juvenile frog. Earlier there is a modifying effect of prolactin in promoting growth and in delaying metamorphosis. Neither the embryonic nor adult form responds to thyroid hormone.

Although the hormonal complement appears to be complete in Salientia amphibians, various intermediate endocrine and metamorphic states have been reported for the other major amphibian order, Caudata, which contains the tailed amphibians, such as salamanders. Thus, the caudates provide many useful examples for the study of comparative endocrinology.

The bullfrog tadpole, lacking thyroid hormone, is the ideal vertebrate for studying the mechanism of thyroid hormone action. In amphibians, thyroid hormone elicits primarily irreversible differentiation responses with no significant reversible calorigenic effects, as observed in mammals. The amphibian response illustrates many of the most significant adaptations that have contributed to vertebrate evolution. Thus far, the data support the T3-nuclear receptor mechanism with new mRNAs producing different proteins in the cytoplasm, thereby initiating differentiation. The tadpole erythrocyte and, to a lesser extent, tail fin cells afford an example of receptor induction in which T3 stimulates up to a fivefold increase in nuclear receptors at metamorphic climax.

Other points of interest covered by Frieden included the limited evidence for cell membrane effects and the substituted thryonine structure which, while favoring iodine and the other halogens, does not require a halogen for significant thyromimetic activity.

Summarized below are 10 advantages of the bullfrog tadpole as a biological model presented by Frieden:

- It is a vertebrate that lacks, and is specifically responsive to, thyroid hormone.

- The endocrine system of the tadpole has the same control components as thyroid hormone in mammalian systems.

- Thyroid hormone induction can be compared to natural spontaneous metamorphosis.

- In contrast to mammalian embryonic forms, tadpoles are free-swimming and independent of maternal environment.

- At the onset of metamorphosis tadpoles stop feeding and produce a self-contained system that is unaffected by dietary constituents.

- A dramatic metabolic contrast is provided by regressing tissues (e.g., gills, tail, and intestine) and developing tissues (e.g., limbs and lungs).

- Some tadpole tissues, such as liver, undergo no significant cell division during metamorphosis. Molecular changes can be determined in a fixed population of cells uncomplicated by the presence of new daughter cells.

- Metamorphosis provides numerous examples of changes in structure and amounts of numerous proteins, including tail hydrolases, changes in enzymes of the urea cycle (liver) and switches in hemoglobins, numerous serum proteins, digestive enzymes, and retinal pigments.

- The tadpole is poikilothermic, and the rates of molecular events in these animals can be accelerated or slowed by temperature control.

- Bullfrog tadpoles (*Rana catesbeiana*) are a convenient size, their stages of metamorphosis are easily described, and they are sensitive to physiological levels of thyroid hormone, which may be given either by injection or addition to the aquatic medium.

REFERENCES

Abdel-Aal, A. I., R. M. Roe, and B. D. Hammock. 1984. Kinetic properties of the inhibition of juvenile hormone esterase by two trifluoromethylketones and O-ethyl,S-phenyl phosphoramidothioate. Pestic. Biochem. Physiol. 21:232-241.

Agui, N., N. A. Granger, L. I. Gilbert, and W. E. Bollenbacher. 1979. Cellular localization of the insect prothoracicotropic hormone: *In vitro* assay of a single neurosecretory cell. Proc. Natl. Acad. Sci. USA 76:5694-5698.

Aschoff, J., ed. 1981. Handbook of Behavioral Neurobiology. Biological Rythms Vol. 4. Plenum Press, New York. 563 pp.

Bargiello, T. A., and M. W. Young. 1984. Molecular genetics of a biological clock in *Drosophila*. Proc. Natl. Acad. Sci. USA 81: 2142-2146.

Bowen, M. F., D. S. Saunders, W. E. Bollenbacher, and L. I. Gilbert. 1984. *In vitro* reprogramming of the photoperiodic clock in an insect *brain-retrocerebral* complex. Proc. Natl. Acad. Sci. USA 81:5881- 5884.

Bullock, T. H. 1981. Implications for neuroethology from comparative neurophysiology. Pp. 53-75 in J.-P. Ewert, R. R. Capranica, and D. J. Ingle, eds. Advances in Vertebrate Neuroethology, Vol. 56, NATO ASI Ser. A, Life Sciences. Plenum Press, New York.

de Vlieger, T. A., K. S. Kits, A. ter Maat, and J. C. Lodder. 1980. Morphology and electrophysiology of the ovulation hormone producing neuro-endocrine cells of the freshwater snail *Lymnaea stagnalis* (L). J. Exp. Biol. 84:259-217.

Dunlap, J. C., and J. W. Hastings. 1981. The biological clock in *Gonyaulax* controls luciferase activity by regulating turnover. J. Biol. Chem. 256:10509-10518.

Duve, H., and A. Thorpe. 1984. Comparative aspects of insect-invertebrate neurohormones. Pp. 171-195 in A. Borkovec and T. Kelly, eds. Insect Neurochemistry and Neurophysiology. Plenum Press, New York.

Evered, D., M. O'Connor, and J. Whelan, eds. 1982. Neuropharmacology of Insects. Ciba Foundation Symposium 88. Ciba Pharmaceutical Company (Medical Education Division), Newark, New Jersey. 330 pp.

Feldman, J. F. 1982. Genetic approaches to circadian clocks. Ann. Rev. Plant Physiol. 33:583-608.

Feldman, J. F., and J. C. Dunlap. 1983. *Neurospora crassa*: A unique system for studying circadian rhythms. Photochem. Photobiol. Rev. 7:319-368.

Gainer, H., ed. 1977. Peptides in Neurobiology. Plenum Press, New York. 464 pp.

Galton, V. A. 1984. Putative nuclear triiodothyronine receptors in tadpole erythrocytes: Regulation of receptor number by thyroid hormone. Endocrinology 114:735-743.

Geraerts, W. P. M., and S. Bohlken. 1976. The control of ovulation in the hermaphroditic freshwater snail *Lymnaea stagnalis* by the neurohormone of the caudodorsal cells. Gen. Comp. Endrocrinol. 28:350-357.

Geraerts, W. P. M., and J. Joosse. 1975. Control of vitellogenesis and of growth of female accessory sex organs by the dorsal body hormone (DBH) in the hermaphroditic freshwater snail *Lymnaea stagnalis*. Gen. Comp. Endocrinol. 27:250-464.

Gilbert, L. I., and E. Frieden, eds. 1981. Metamorphosis: A Problem in Developmental Biology. Plenum Press, New York. 578 pp.

Gilbert, L. I., W. E. Bollenbacher, W. Goodman, S. L. Smith, S. Agui, N. Granger, and B. J. Sedlak. 1980. Hormones controlling insect metamorphosis. Rec. Prog. Hormone Res. 36:401-449.

Gilbert, L. I., W. E. Bollenbacher, N. Agui, N. A. Granger, B. J. Sedlack, D. Gibbs, and C. M. Buys. 1981. The prothoracicotropes: Source of the prothoracicotropic hormone. Amer. Zool. 21:641-653.

Gottfried, H., and R. I. Dorfman. 1970. Steroids of invertebrates IV. On the optic tentacle-gonadal axis in the control of the male-phase ovotestis in the slug (*Ariolimax californicus*). Gen. Comp. Endocrinol. 15:101-119.

Hammock, B. D. In press. Regulation of juvenile hormone titer: degradation. In G. Kerkut and L. Gilbert, eds. Comprehensive Insect Physiology, Biochemistry and Pharmacology, Vol. 7. Pergamon Press, New York.

Haynes, L. W. 1980. Peptide neuroregulators in invertebrates. Prog. Neurobiol. 15:205-245.

Hildebrand, J. G. In press. Metamorphosis of the insect nervous system: Influences of the periphery on the postembryonic development of the antennal sensory pathway in the brain of *Manduca sexta*. A. I. Selverston, ed. Model Neural Networks and Behavior. Plenum Press, New York.

Hildebrand, J. G., L. M. Hall, and B. C. Osmond. 1979. Distribution of binding sites for ^{125}I-labeled oc-bungarotoxin in normal and deafferented antennal lobes of *Manduca sexta*. Proc. Natl. Acad. Sci. USA 76:499-503.

Hildebrand, J. G., S. G. Matsumoto, L. P. Tolbert, A. M. Schneiderman, and S. M. Camazine. 1982. Postembryonic development of the antennal lobes in the moth *Manduca sexta*. Neurosciences Res. Prog. Bull. 20:891-900.

Jacklet, J. W. 1981. Circadian timing by endogenous oscillators in the nervous system: Toward cellular mechanisms. Biol. Bull. 160:199-227.

Johnson, C. H., J. F. Roeber, and J. W. Hastings. 1984. Circadian changes in enzyme concentration account for rhythm of enzyme activity in *Gonyaulax*. Science 223:1428-1430.

Joosse, J. 1984. Photoperiodicity, rhythmicity, and endocrinology of reproduction in the snail *Lymnaea stagnalis*. Pp. 204-220 in R. Porter and G. M. Collins, eds. Photoperiodic Regulation of Insect and Molluscan Hormones. Ciba Foundation Symposium 104. Ciba Pharmaceutical Company (Medical Education Division), Newark, New Jersey.

Kits, K. S. 1980. States of excitability in ovulation hormone producing neuroendocrine cells of *Lymnaea stagnalis* (Gastropoda) and their relation to the egg-laying cycle. J. Neurobiol. 11:397-410.

Kramer, K. In press. Vertebrate hormones in insects. In G. Kerkut and L. I. Gilbert, eds. Comprehensive Insect Physiology, Biochemistry and Pharmacology, Vol. 7. Pergamon Press, New York.

Kravitz, E. A., B. S. Beltz, S. Glueman, M. F. Goy, R. M. Harris-Warrick, M. R. Johnson, M. S. Livingstone, T. L. Schwartz, and K. K. Swicki. 1983. Neurohormones and lobsters: Biochemistry to behavior. Trends in Neurosci. 6:346-349.

Kyriacou, C. P., and J. C. Hall. 1980. Circadian rhythm mutations in *Drosophila melanogaster* affect short-term fluctuations in the males' courtship song. Proc. Natl. Acad. Sci. USA 77:6729-6733.

Matsumoto, S. G., and J. G. Hildebrand. 1981. Olfactory mechanisms in the moth *Manduca sexta*: Response characteristics and morphology of central neurons in the antennal lobes. Proc. R. Soc. Lond., Ser. B 213:249-277.

Maxwell, G. D., and J. G. Hildebrand. 1981. Anatomical and neurochemical consequences of deafferentation in the development of the visual system of the moth *Manduca sexta*. J. Comp. Neurol. 195:667-680.

McCrone, E. J., and P. G. Sokolove. 1979. Brain-gonad axis and photoperiodically-stimulated sexual maturation in the slug, *Limax maximus*. J. Comp. Physiol. 133:117-123.

McCrone, E. J., J. van Minnen, and P. G. Sokolove. 1981. Slug reproductive maturation hormone: *In vitro* evidence for long-day stimulation of secretion from brains and cerebral ganglia. J. Comp. Physiol. 143:311-315.

Ota, K., and B. D. Hammock. 1980. Cytosolic and microsomal epoxide hydrolases: Differential properties in mammalian liver. Science 207:1479-1481.

Pittendrigh, C. S., J. Elliott, and T. Takamura. 1984. The circadian component in photperiodic induction. Pp. 26-47 in R. Porter and G. M. Collins, eds. Photoperiodic Regulation of Insect and Molluscan Hormones. Ciba Foundation Symposium 104. Ciba Pharmaceutical Company (Medical Education Division), Newark, New Jersey.

Porter, R., and G. M. Collins, eds. 1984. Photoperiodic Regulation of Insect and Molluscan Hormones. Ciba Foundation Symposium 104. Ciba Pharmaceutical Company (Medical Education Division), Newark, New Jersey. 298 pp.

Reddy, P., W. A. Zehring, D. A. Wheeler, V. Pirvotta, C. Hadfield, J. C. Hall, and M. Rosbash. 1984. Molecular analysis of the period locus in _Drosophila melanogaster_ and identification of a transcript involved in biological rhythms. Cell 38:701-710.

Sanes, J. R., and J. G. Hildebrand. 1976. Origin and morphogenesis of sensory neurons in an insect antenna. Dev. Biol. 51:300-319.

Sanes, J. R., J. G. Hildebrand, and D. J. Prescott. 1976. Differentiation of insect sensory neurons in the absence of their normal synaptic targets. Dev. Biol. 52:121-127.

Scharrer, B. 1976. Neurosecretion -- comparative and evolutionary aspects. Prog. Brain Res. 45:125-137.

Scharrer, B. 1978a. Peptidergic neurons: Facts and trends. Gen. Comp. Endocr. 34:50-62.

Scharrer, B. 1978b. An Evolutionary Interpretation of the Phenomenon of Neurosecretion. The James Arthur Lecture on the Evolution of the Human Brain, No. 47. The American Museum of Natural History, New York. 17 pp.

Scharrer, B. 1981. Neuroendocrinology and histochemistry. Pp. 11-20 in P. J. Stoward and J. M. Polak, eds. The Widening Horizons. John Wiley & Sons, Ltd., Chichester, England.

Schneiderman, A. M., S. G. Matsumoto, and J. G. Hildebrand. 1982. Transsexually grafted antennae influence development of sexually dimorphic neurones in moth brain. Nature 298:844-846.

Schuler, R. L., B. D. Hardin, and R. W. Niemeier. 1982. _Drosophila_ as a tool for the rapid assessment of chemicals for teratogenicity. Terat. Carc. Mutag. 2:293-301.

Schwartz, L. M., and J. W. Truman. 1983. Hormonal control of the rates of metamorphic development in the tobacco hornworm *Manduca sexta*. Dev. Biol. 99:103-114.

Sokolove, P. G., G. R. Melrose, T. M. Gordon, and M. C. O'Neill. 1983. Stimulation of spermatogonial DNA synthesis in slug gonad by a factor released from cerebral ganglia under the influence of long days. Gen. Comp. Endocrinol. 50:95-104.

Sokolove, P. G., E. J. McCrone, J. van Minnen, and W. C. Duncan. 1984. Reproductive endocrinology and photoperiodism in a terrestrial slug. Pp. 189-203 in R. Porter and G. M. Collins, eds. Photoperiodic Regulation of Insect and Molluscan Hormones. Ciba Foundation Symposium 104. Ciba Pharmaceutical Company (Medical Education Division), Newark, New Jersey.

Stefano, G. B., and B. Scharrer. 1981. High affinity binding of an enkephalin analog in the cerebral ganglion of the insect *Leucophaea maderae* (Blattaria). Brain Res. 225:107-114.

Stefano, G. B., B. Scharrer, and P. Assanah. 1982. Demonstration, characterization and localization of opioid binding sites in the midgut of the insect *Leucophaea maderae* (Blattaria). Brain Res. 252:205-212.

Strausfield, N. J., and M. E. Adams. 1983. Functional Neuroanatomy. Springer-Verlag, New York. 427 pp.

Takeda, N. 1983. Endocrine regulation of reproduction in the snail (Euhadra peliomphala). Pp. 106-111 in J. Lever and H. H. Boer, eds. International MiniSymposium of Molluscan Endocrinology. Elsevier/North Holland, Amsterdam, the Netherlands.

Truman, J. W. 1984. Physiological aspects of the two oscillators that regulate the timing of eclosion in moths. Pp. 221-239 in R. Porter and G. M. Collins, eds. Photoperiodic Regulation of Insect and Molluscan Hormones. Ciba Foundation Symposium 104. Ciba Pharmaceutical Company (Medical Education Division), Newark, New Jersey.

Truman, J. W., P. H. Taghert, P. F. Copenhaver, N. J. Tublitz, and L. M. Schwartz. 1981. Eclosion hormone may control all ecdyses in insects. Nature 291:70-71.

Walton, B. T. 1983. Use of the cricket embryo (*Acheta domesticus*) as an invertebrate teratology model. Fundam. Appl. Toxicol. 3:233-236.

Woolum, J. C., and F. Strumwasser. 1980. The differential effects of ionizing radiation on the circadian oscillator and other functions in the eye of *Aplysia*. Proc. Natl. Acad. Sci. USA 77:5542-5546.

Yoshizato, K., and E. Frieden. 1975. Increase in binding capacity for triiodothyronine in tadpole tail nuclei and during metamorphosis. Nature 254:705-707.

Appendix E

Models for the Study of Diseases and Aging

Workshop Program
June 27, 1984
Woods Hole, Mass.
Vincent Cristofalo and Samuel Ward
Organizers and Chairmen

Paul Gross Marine Biological Laboratory Woods Hole, Mass.	Introductory Talk: Lower Animals and Biomedical Science
Gerald Weissman New York University Medical Center New York, N.Y.	Marine Sponges as Models of Inflammation
Daniel Drachman The Johns Hopkins School of Medicine Baltimore, Md.	Myasthenia Gravis as a Model for Human Autoimmune Disorders
George Cahill Howard Hughes Medical Foundation Boston, Mass.	Diabetes: Crossroads of Genetics, Immunology, and Environment
Leon Glass McGill University Montreal, Quebec, Canada	Oscillation Chaos and Cardiac Arrythmias
Jay Seegmiller University of California, San Diego La Jolla, Calif.	Possible Genetic Models for the Study of Human Aging
Samuel Goldstein University of Arkansas Little Rock, Ark.	The Cultured Human Fibroblast: A Model for the Study of Cellular and Molecular Aging
Russell Ross University of Washington Seattle, Wash.	Atherosclerosis: A Response to Injury Gone Awry
Charles Stiles Dana Farber Cancer Research Institute Boston, Mass.	The Role of Oncogenes in the Biology of Normal Cells: Insights from Platelet-Derived Growth Factor

Workshop Summary

In the fifth workshop, the participants examined the different kinds of models that have contributed to our understanding of specific diseases and aging. No attempt was made to be comprehensive. Instead, the presentations were focused on several specific diseases and medical problems, including inflammatory responses, myasthenia gravis, diabetes, atherosclerosis, cardiac arrhythmia, cancer, and aging. The speakers described the roles that both invertebrate and vertebrate models have played in furthering understanding in their research areas, and they discussed both the advantages and limitations of the different models.

Paul Gross (Marine Biological Laboratory) opened the workshop with the following definition of a model in biomedical science: a system in which all, or most, components of a normal or pathological process of human biology are present but which facilitates study of the process because it is simpler overall, it is cheaper or easier to use than human subjects, or it is available in quantities sufficient to ensure that random variation does not influence the results of study in an unknown way. He pointed out that in principle, the biochemical unity of biological processes means that there are many potential models. Medical researchers select the one that they believe is the best emulator of the essential features of a disease. The model must then be analyzed and its validity tested.

Many new diseases have recently been discovered in marine animals through careful pathological examination of specimens collected for the Marine Biological Laboratory. Not enough is known about most of these marine organisms to be able to determine their usefulness as specific models for diseases, but their potential is substantial. Gross used biological effects of ultrasound as an example of a potential medical problem that is in need of good models. Ultrasonic imaging is being used routinely to monitor pregnancy, but, he argued, not enough is known about its potential effects on developing embryos to be certain it is really safe. Marine embryos, whose development has been studied for more than a century, should offer good biological models for this purpose.

Gross also pointed out the important difference between models and modelling. He described modelling as the search for the variables underlying a physiological process and the analysis of their quantitative importance. This is distinct from the search for models in biomedical research, which is the quest for adequate substitute systems for a disease.

Gerald Weissman (New York University Medical Center) described the role of neutrophils in acute immunologic inflammatory responses such as those found in arthritis. The neutrophil responds first by moving toward a chemoattractant released either by bacteria or by complement. It then becomes sticky, attaching to, aggregating on, and eventually infiltrating tissue. This response is one of many examples of stimulus response coupling by cells. As such it can be studied in cultured human cells, but it can also be studied in dissociated cells of the sponge *Microciona intestinalis* (Rich *et al.*, 1984). These cells can be obtained in large number by disaggregation of a sponge in calcium-free seawater. Reaggregation, which can be easily measured by light scattering, can then be studied as a model of neutrophil aggregation.

The aggregation of sponge cells is inhibited by most nonsteroidal antiinflammatory drugs (NSAIDs), and the resulting dose-response curve is very similar to that for the effect of NSAIDs on neutrophils. However, the sponge cells are resistant to nearly all prostaglandins except leucotriene 4. This implies that NSAIDs can affect cell aggregation without mediation by effects on prostaglandins. Direct measurements show that NSAIDs act directly by inhibiting calcium ion uptake, probably into a membrane-associated pool.

This example illustrates that invertebrate cells can be utilized effectively to help understand how important pharmacological drugs act. These phenomena are being investigated at a subcellular level by assay of specific enzyme activities and measurement of ion pools and transport. Enough is known from studies of human cells not only to recognize the limitations of the sponge cell model but also to define the similarities of its inflammatory responses to those of humans, thereby exploiting the experimental advantages of this easily obtained, large homogeneous population of cells. However, it is precisely because the cells differ in the enzymes of prostaglandin biosynthesis that the similar response to NSAID is so interesting; this difference suggests that these drugs can act directly on calcium ion levels in the cell.

Daniel Drachman (The Johns Hopkins School of Medicine) described myasthenia gravis (MG) -- a human neuromuscular disorder characterized by weakness and fatigue of skeletal muscles. The basic defect causing the symptoms is a reduction of acetylcholine receptors (ACHRs) at neuromuscular junctions, which is brought about by a humoral autoimmune attack. Six different models have been useful in elucidating the pathogenesis of MG (reviewed in Drachman, 1981, 1983).

One is the human disease itself, which may serve as a model for understanding other autoimmune diseases. In the 1930s, anticholinesterase agents were found to benefit MG patients. This suggested that the neuromuscular junctions might be the site of the primary defect in MG. Subsequent experiments with the neurotoxin alpha-bungarotoxin, isolated from the venom of the elapid snake *Bungarus multicinctus*, demonstrated that myasthenic patients had a striking deficit of AChRs at the neuromuscular junction. This neurotoxin had been shown to be specific for AChRs in earlier studies on chicken muscles in culture, which had established its specificity and developed quantitative methods for measuring binding.

To determine if a deficit in receptors could explain the physiological symptoms, the rat was developed as a model by giving it cobra neurotoxin, which blocks AChRs. This produced typical myasthenic symptoms that improved after the administration of anticholinesterase drugs. Thus a deficit in AChRs could be responsible for the symptoms of MG. The validity of this conclusion depends on how similar the neuromuscular physiology of the rat is to humans. If nonmammalian organisms had been used, there would be less confidence that the results would be valid for humans.

Indirect evidence has suggested that MG might be an autoimmune disease. To test this, an experimental autoimmune MG model was developed. It was found that immunization of several animal species with AChR purified from invertebrates would induce myasthenic symptoms. Although the origin of antibodies in this model differs from that of the human disease, the disease symptoms could be produced by autoimmune attack. This model has also enabled investigators to test immunosuppressive agents for potential prevention of the disease.

If MG were caused by an autoimmune response to AChRs, then patients should have antibodies to AChRs in their blood. This was found to be true in general, but the antibody titer did not correlate well with the severity of symptoms. To investigate the role of these antibodies further, a passive transfer model of the disease was developed by injecting an MG patient's sera into mice. The mice developed MG symptoms, giving further confirmation of the autoimmune basis for MG. This passive transfer model has been useful for investigating other autoimmune diseases as well. Its success depends on immunological cross-reaction between the antigens of the experimental animals, so it is most likely to be useful in animals most closely related to human beings.

Two cell culture models have also been useful for studying MG. One is a rat skeletal muscle cell culture, which has made possible the determination of the effect of antibodies on the AChR. It was found that the IgG from 90% of MG patients increased the turnover of AChR in the muscle cell membranes. This was independent of cross-linking of the AchR in the membrane. This increased turnover has proved to be a general phenomenon of much interest in cell biology. Another technique being tried is the development of suppressor lymphocyte cultures, which may provide a future method of immunotherapy for MG and other autoimmune diseases.

Our understanding of the molecular basis of MG illustrates particularly clearly the essential role that both animal models and cell cultures must play in medical research.

George Cahill (Howard Hughes Medical Foundation) described the epidemiology and pathogenesis of juvenile-onset diabetes and the usefulness of the Wistar BB rat as an animal model of the disease (Skylar and Cahill, 1981). In the United States, juvenile-onset diabetes occurs in approximately 1 of every 300 people. The onset of this disease peaks in children about 12 years of age. The substantial genetic component in the disease is reflected in the 25% to 50% concordance between identical twins (Barnett _et al._, 1981; Srikanta _et al._, 1983). There is also a correlation with histocompatability antigens D3 and D4. Like MG, juvenile-onset diabetes appears to be an autoimmune disease. What initiates the disease is not known. It could be a virus. It is known, however, that the disease progresses after antibodies are produced against a 64Kd membrane protein found in pancreatic beta cells. This leads to autoimmune destruction of the beta cells and, thus, loss of the ability to produce insulin. The cells are not replaced, and remissions of the disease have not been reported.

In 1974 the Wistar BB rat strain was found to develop symptoms similar to juvenile-onset diabetes at about 60 to 120 days of age (Marliss _et al._, 1982). The pathology of the disease in rats is almost identical to that of the human form. For example, the disease can be passively transferred by blood serum, and antibodies to the beta cell 64Kd protein in humans cross-react with a similar beta cell protein in the rat (Dryberg _et al._, 1982).

The disease can be prevented by interfering with the immune system before the onset of the disease. For example, cyclosporin treatment before onset can prevent the disease, but it has no effect after onset. A similar therapy might work in humans, but it is too expensive at present to identify all those at risk. Moreover, the side effects of immunosuppression are severe.

Further understanding of the pathogenesis of the diabetes would be facilitated by study of a cell culture model of beta cells, but this has not yet been achieved.

Leon Glass (McGill University) described mathematical modelling of biological oscillators using cardiac electrophysiology as an example. He proposed that many human diseases can be regarded as dynamic diseases whose physiological characteristics become altered causing a qualitative change in behavior of the system known as a bifurcation (Glass and Mackey, 1979; Mackey and Glass, 1977). By mathematical modelling of the physiological system, one gains a better understanding of the relative importance of the different characteristics and, thus, which ones to try to manipulate to cure the disease.

One biological model that can be used to test and develop mathematical models of cardiac arrythmias is the spontaneously beating cultured ventricular cell from the embryonic chicken heart. The onset of arrhymthias in the beating pattern in cell culture matched the appearance of bifurcations in a mathematical model (Glass *et al*., 1984; Guevara *et al*., 1981).

Discussion of this presentation revealed uncertainty about the general usefulness of such models. Some argued that the apparent generality of such models might mask fundamentally different mechanisms for generating oscillations. Since diseases act by altering normal physiological activity in a specific system, knowing the specific molecular basis for oscillations would be an alternative to general mathematical description of their possible behavior. Glass suggested that the committee should consider additional examples of mathematical modelling. He stated that there is a need for more research collaboration between mathematical modellers and the users of biological models. (A sixth workshop on mathematical modelling was convened because the committee believed that the topic had not been sufficiently addressed in the other workshops.)

Jay Seegmiller (University of California, San Diego) reviewed some of the important studies of specific genetic diseases used as approaches to understanding normal human gene function, including onset of senesence (Lust *et al*., 1981; Seegmiller, 1972, 1977, 1980a,b). All the work was done in human cell cultures, either lymphoblastoid or fibrolast-like cells. Seegmiller cited some of his own work and that of others to elucidate the deficiency of the enzyme hypoxathine guanine phosphoribosyl transferase in Lesch Nyhan disease. He also spoke about the disease chondrocalcinosis, which is characterized by the abnormal deposition of calcium pyrophosphate crystals in afflicted individuals. The disease is inherited as an autosomal dominant. Studies of cultured lymphoblasts and fibroblasts from afflicted individuals have demonstrated that the cells contained abnormally high amounts of calcium pyrophosphate. Since sarcoidosis, the most common form of arthritis in the elderly, is marked by significant elevations of calcium pyrophosphate in cells and in lymph nodes, chondrocalcinosis might serve as a probe in understanding at least one of the major diseases of aging.

Seegmiller described several genetic diseases in humans that, in some aspects, appear to result in precocious aging. These diseases include the Hutchinson Gilford syndrome (progeria), Werner's syndrome, Down's syndrome, Cockayne's syndrome, Bloom's syndrome, and others. The afflicted individuals have some, but not all, features of aging.

Models for these genetically determined diseases are used at two different levels. At one level, fibroblast and lymphoblast cultures used as models to study the biochemical basis of human disease have been very productive in advancing our understanding of human disease. At a second level, human diseases, especially genetically determined disease, can be used as models or probes for understanding normal gene function.

Samuel Goldstein (University of Arkansas) reviewed cultured human fibroblasts as a model to study biological aging and age-dependent diseases. He emphasized their life history and the analogy between the growth properties of these cells and various cell types *in vivo*. The decline in replicative capacity of these cultured human fibroblasts bears some relationship to aging and, thus, might be useful in the study of the mechanism of aging in humans (Goldstein, 1979).

It is likely that aging is not due to error in protein synthesis but, rather, that it has a genetic basis. Goldstein and his coworkers have been interested in the highly repetitious DNA, whose function is not understood but is believed to be related to regulation. This is in contrast to the unique DNA sequences that code for proteins. Goldstein has found that human fibroblasts lose repetitious DNA as they traverse their limited replicative life spans. Qualitative changes in the organization of these sequences may be even more important. After subculturing, fibroblasts accumulate increased amounts of covalently closed circular DNA (cccDNA), first detected by a specific "Inter-Alu" probe. This low copy DNA sequence was subcloned from a 15-kb fragment (derived from Maniatis' human genome library) containing a rich concentration of the putatively mobile Alu family of dispersed repetitious DNAs (approxmiately 5 x 10^5/haploid genome). Goldstein's group has now isolated total cccDNAs nonselectively and has cloned them into the plasmid vector pBR322.

Preliminary characterization of 107 cloned cccDNAs has revealed the remarkable coexistence of Alu and Kpn sequences, the latter representing another family of dispersed and repetitious DNAs (approximately 5 x 10^4/haploid genome). This suggests that juxtaposition of these two sequence families, or a specific subclass of each, endows DNA with mobility to the extent that these segments are excised from chromosomal DNA into cccDNAs with the potentional to reinsert at novel chromosomal locations and to subvert normal resident genes into aberrant acts of expression or replication (Goldstein, 1979; Goldstein and Shmookler-Reis, 1984).

More recently, Goldstein and his colleagues have shown that the c-Ha-ras protooncogene is amplified up to fourfold in seven consecutively examined strains of late-passage fibroblasts and that this occurred irrespective of donor age. Moreover, c-Ha-ras gene amplification was accompanied by a commensurate increase in the specific mRNA and its protein product, p21, which shows close homology to the guanosine triphosphate (GTP) binding subunit of adenyl cyclase. Thus, amplification and overexpression of the c-Ha-ras protooncogene occurs during protracted replication of *normal* cells, which may help to explain the increasing risk to malignant transformation seen as a function of age.

Therefore, human cells in culture represent a useful model system for approaching the mechanism of human aging at the cellular level. Since aging is probably genetically controlled and may involve a process of gene switching, regulatory factors in DNA that control gene expression are of interest. By using human cells in culture, we can learn about the human genome, the regulation of gene expression, and the process of human aging.

Russell Ross (University of Washington, Seattle) reviewed the injury theory of atherosclerosis. He proposed that damage to the vascular endothelium from a variety of causes, including hypercholesterolemia and hypertension, causes a lesion involving smooth muscle cells, macrophages, and lipid accumulation. Endothelial cells form a thromboresistant barrier because of their surface coat of heparin sulfate and their capacity to synthesize prostacyclin. These cells can metabolize vasoactive substances and can act as a permeability barrier to plasma constituents. Varying forms of injury may cause either major or subtle alterations in any of these functional characteristics, sometimes with no visible morphologic change in the endothelial cells. At the other extreme, if the injury is sufficiently severe, enthothelial cells may detach from each other and, possibly, from the connective tissue. If the latter were to take place at sites where blood flow is altered by bifuractions in the artery, the cells may desquamate into the lumen of the vessel, exposing the subendothelial connective tissue (Brown and Goldstein, 1984; Faggiotto and Ross, 1984; Faggiotto *et al.*, 1984; Ross, 1983; Schwartz and Ross, 1984).

The hypothesis suggests that if either the nonthrombogenic character of the endothelium is altered or endothelial desquamation occurs, platelets and monocytes could interact and perhaps attach to the artery wall. If the monocytes attach and enter the tissue, they can develop into macrophages. Macrophages can synthesize, and both platelets and macrophages contain, a number of potent substances that can markedly alter the metabolism of the smooth muscle cells of the artery wall, potentially resulting in profound consequences. In particular, the platelet-derived growth factor (PDGF) and the macrophage-derived growth factor (MDGF) are two potent mitogens that have numerous metabolic effects on susceptible cells, such as smooth muscle and fibroblasts.

The response-to-injury hypothesis suggests that these two growth factors would be included among substances that may be released into the tissue following endothelial injury and interaction of platelets and monocytes or macrophages. They could chemotactically attract smooth muscle cells from the media of the artery into the intima and stimulate their proliferation, leading to the development of an intimal, smooth muscle, preatherosclerotic, proliferative lesion. The hypothesis suggests that, if the injury is brief and nonrecurrent, the proliferative response may regress and the lesion would therefore be reversible. On the other hand, if the injury is chronic, lasting over many years, and takes place during, or is due to, hypercholesterolemia, the lesions may progress and could become laden with lipids, resulting in classical lesions of atherosclerosis.

Ross and his colleagues have studied 40 monkeys made hypercholesterolemic with a high fat, high cholesterol diet, so that their serum cholesterol levels were 600 to 1,000 ng/ml. This is a level equivalent to that of humans with familial hypercholesterolemia. The investigators studied endothelial turnover in the animals by autoradiographic techniques and its morphology by light microscopy and by scanning and transmission electronmicroscopy. After 1 month on the diet, macrophages appeared in the subendothelial space of the monkeys in the form of foam cells (i.e., cells full of lipid droplets).

At 4 months, the first discernible changes in the endothelium could be observed. These were openings between junctional complexes of the endothelial cells overlying the macrophages, exposing the surface of the macrophage and its lamellipodia to the lumen of the artery. In addition, the scanning electron microscope revealed small aggregates of platelets, suggesting that there was exposed subendothelium underneath these platelet aggregates (Faggiotto and Ross, 1984; Faggiotto _et al_., 1984). After 5 months, drastic changes were observed, particularly in the abdominal aorta and iliac and superficial femoral arteries.

Examination of the foam cell lesions by light and electron microscopy revealed that small numbers of smooth muscle cells had accumulated under the macrophages after 2 and 3 months of the dietary regimen and that greater numbers had appeared after 5 months.

Thus, in the hypercholesterolemic nonhuman primate, the first event to occur appears to be the attraction of monocytes from the blood into the subendothelial space. The monocytes may then transform into foam cells, which may possibly have a deleterious effect on the overlying endothelium. Within 5 months, the cumulative effects of chronic hypercholesterolemia upon the endothelium result in separation of endothelial cells and in endothelial denudation, causing platelet adherence, aggregation, and degranulation during the formation of mural thrombi. These platelet and macrophage interactions appear to be associated with increased smooth muscle migration and proliferation, leading to the formation of classic atherosclerosis in these animals.

Ross has also studied the interaction between PDGF and smooth muscle cells (Bowen-Pope and Ross, 1984). Apparently PDGF has a chemotactic effect, which causes the smooth muscle cells to migrate into the lesion, as well as a mitogenic effect. Simply stated, Ross began with a human disease that has been poorly understood until recently, although it is responsible for 50% of the deaths in the United States. By studying samples of arteries from human bypass surgery, he and his coworkers were able to propose a hypothesis. To test the hypothesis they used monkeys as models to study the etiology and progression of the disease. Monkey and human cell cultures were then used as models to elucidate the cellular basis for the lesions. This sequence illustrates the fundamental point that models exist only in terms of the question being asked. For some questions, cellular or even subcellular models are appropriate. For others, when systems physiology is involved, intact experimental animals must be used. Ross used cells, intact experimental animals, and humans to develop an understanding of this disease.

Charles Stiles (Dana Farber Cancer Research Institute) explained that those studying tumor biology fall into two camps. One group believes that cancer has a genetic component and that a discrete set of genes can cause cancer when they are activated. The corollary of this is that the control of those genes has a role in normal biology as well. The second group of researchers approach tumor cell biology from biochemistry and cell biology, especially in their use of growth factors as probes of tumor growth control. Cell growth is regulated by hormones and growth factors. Tumor cells in general have diminished regulation. Various workers have shown that growth factors are very highly conserved in an evolutionary sense. In evolutionary sequence, for example, the appearance of PDGFs precedes the appearance of the blood platelet. The question that Stiles addresses is, "How does this growth factor work?" (Armelin et al., 1984; Cochran et al., 1983; Kelly et al., 1983; Stiles, 1983).

Human PDGF is contained in platelets, and clotted blood serum contains approximately 15 to 20 ng of PDGF per milliliter. Smooth muscle cells, fibroblasts, and glial cells require PDGF for optimum growth, whereas epithelioid cells do not. In addition to its role in stimulating mesenchymal cell proliferation, PDGF has served as an important probe in advancing our understanding of gene expression and tumor cell growth.

There is now an emerging body of evidence that growth factors and oncogenes affect each other. It takes a gene to synthesize a growth factor. At the same time, growth factors regulate gene expression. PDGF may have at least three functions related to cellular oncogenes. In the first, oncogenes may cause the production of a growth factor. For example, simian sarcoma virus appears to have acquired its transforming gene (V-sis) from the structural gene for PDGF. At a second level, the receptor of PDGF is associated with a tyrosine-specific protein kinase, which is an enzyme encoded by several viral oncogenes. At a third level, the role of growth factors lies within the control of genes active early in the cell cycle. To illustrate this third category, Stiles discussed studies demonstrating that from 0.1% to 1.3% of the genes expressed in 3T3 cells are regulated by PDGF. This means that there are between 10 and 30 PDGF-inducible genes. Of these, 5 to 14 genes correspond to low abundance mRNAs in quiescent cells. However, within 1 hour following stimulation with PDGF, these mRNAs increase approximately 20-fold. By 12 hours, they drop considerably. Although the specific role of these genes in early cell cycle events is not known, there is presumably a connection.

Here again the overall thrust of the work shows that mammalian cells in culture, mouse cells in this case, can be used to probe the basic mechanisms that underlie gene expression as well as the controlling factors in tumor cells that provide the basis for the loss of growth regulation.

REFERENCES

Armelin, H. A., M. C. S. Armelin, K. Kelly, T. Stewart, P. Leder, B. H. Cochran, and C. D. Stiles. 1984. Functional role of c-myc in mitogenic response to platelet-derived growth factor. Nature 310:655-660.

Barnett, A. H., C. Eff, R. D. G. Leslie, and D. A. Pyke. 1981. Diabetes in identical twins: A study of 200 pairs. Diabetologia 20:87-93.

Bowen-Pope, D. F., and R. Ross. 1984. Platelet-derived growth factor. Clin. Endocrinol. Metab.: Tissue Growth Factors 13:191-205.

Brown, M. S., and J. L. Goldstein. 1984. How LDL receptors influence cholesterol and atherosclerosis. Sci. Amer. 251:58-66.

Cochran, B. H., A. C. Reffel, and C. D. Stiles. 1983. Molecular cloning of gene sequences regulated by platelet-derived growth factor. Cell 33:939-947.

Drachman, D. B. 1981. The biology of Myasthenia gravis. Annu. Rev. Neurosci. 4:195-225.

Drachman, D. B. 1983. Myasthenia gravis: Immunobiology of a receptor disorder. Trends Neurosci. 6:446-451.

Dryberg, T. D., A. F. Nakhooda, S. Baekkeskov, A. Lernmark, P. Poussier, and E. B. Marliss. 1982. Islet cell surface antibodies and lymphocyte antibodies in the spontaneously diabetic BB Wistar rat. Diabetes 31:278-281.

Faggiotto, A., and R. Ross. 1984. Studies of hypercholesterolemia in the nonhuman primate. II. Fatty streak conversion to fibrous plaque. Arteriosclerosis 4:341-356.

Faggiotto, A., R. Ross, and L. Harker. 1984. Studies of hypercholesterolemia in the nonhuman primate. I. Changes that lead to fatty streak formation. Arteriosclerosis 4:323-340.

Glass, L., and M. C. Mackey. 1979. Pathological conditions resulting from instabilities in physiological control systems. Ann. N. Y. Acad. Sci. 316:214-235.

Glass, L., M. R. Guevara, J. Belair, and A. Shrier. 1984. Global bifurcations of a periodically forced biological oscillators. Phys. Rev. A29:1348-1357.

Goldstein, S. 1979. Senescence. Pp. 2001-2025 in L. J. DeGroot, G. F. Cahill, Jr., W. D. Odell, L. Martini, J. T. Potts, Jr., D. H. Nelson, E. Steinberger, and A. I. Winegrad, eds. Endocrinology, Vol. III. Grune and Stratton, New York.

Goldstein, S., and R. Shmookler-Reis. 1984. Genetic modifications during cellular aging. Mol. Cell. Biochem. 64:15-30.

Guevara, M. R., L. Glass, and A. Shrier. 1981. Phase locking, period-doubling bifurcations, and irregular dynamics in periodically stimulated cardiac cells. Science 214:1350-1352.

Kelly, K., B. H. Cochran, C. D. Stiles, and P. Leder. 1983. Cell-specific regulation of the c-myc gene by lymphocyte mitogens and platelet-derived growth factor. Cell 35:603-610.

Lust, G., G. Farure, P. Netter, and J. E. Seegmiller. 1981. Increased pyrophosphate in fibroblasts and lymphoblasts from patients with hereditary diffuse articular chondrocalcinosis. Science 214:809-810.

Mackey, M. C., and L. Glass. 1977. Oscillation and chaos in physiological control systems. Science 197:287-289.

Marliss, E. B., A. F. Nakhooda, P. Poussier, and A. A. F. Sima. 1982. The diabetic syndrome of the "BB" Wistar rat: Possible relevance to type I (insulin-dependent) diabetes in man. Diabetologia 22:225-232.

Rich, A. M., G. Weissman, C. Anderson, L. Vosshall, K. A. Haines, T. Humphreys, and P. Dunham. 1984. Calcium dependent aggregation of marine sponge cells is provoked by leukotriene B_4 and inhibited by inhibitors of arachidonic acid oxidation. Biochem. Biophys. Res. Commun. 121:863-870.

Ross, R. 1983. Atherosclerosis: A question of endothelial integrity and growth control of smooth muscle. Harvey Lectures, Ser. 77:161-182.

Schwartz, S., and R. Ross. 1984. Cellular proliferation in atherosclerosis and hypertension. Prog. Cardiovasc. Dis. 26:355-372.

Seegmiller, J. E. 1972. New prospects for understanding and control of genetic diseases. Arch. Intern. Med. 130:181-185.

Seegmiller, J. E. 1977. Newer developments in tissue culture: Further aid in the study of inborn errors of metabolism. Pp. 187-198 in Proceedings of the Annual Meeting of the Society for the Study of Inborn Errors of Metabolism (July 1976, Edinburgh, Scotland). Medical and Technical Publishing Co., Lancaster, England.

Seegmiller, J. E. 1980a. Heberden oration 1979: Human aberrations of purine metabolism and their significance for rheumatology. Ann. Rheum. Dis. 39:103-117.

Seegmiller, J. E. 1980b. Diseases of purine and pyrimidine metabolism. Pp. 777-937 in P. K. Bondy and L. E. Rosenberg, eds. Metabolic Control and Disease, 8th ed. W. B. Saunders, Philadelphia.

Skylar, J. S., and G. F. Cahill, Jr. 1981. Symposium on diabetes mellitus. Parts I-III. Am. J. Med. 70:101-214; 325-378; 579-630.

Srikanta, S., O. P. Ganda, R. A. Jackson, R. E. Gleason, A. Kaldany, M. R. Garovoy, E. L. Milford, C. B. Carpenter, J. S. Soeldner, and G. S. Eisenbarth. 1983. Type I diabetes mellitus in monozygotic twins: Chronic progressive beta cell dysfunction. Ann. Intern. Med. 99:320-326.

Stiles, C. D. 1983. The molecular biology of platelet-derived growth factor. Cell 33:653-655.

Appendix F

Models for the Study of Development

Workshop Program

August 16, 1984
Woods Hole, Mass.
George Streisinger, Organizer
William Wood, Chairman
J. Woodland Hastings, Rapporteur

William B. Wood University of Colorado Boulder, Colo.	Introductory Talk: Polarization During First Cleavage in the Nematode
Paul Sternberg University of California, San Francisco San Francisco, Calif.	The Role of Binary Switches in Nematode Development
Gary Struhl Harvard University Cambridge, Mass.	Selector Genes That Regulate Development Fate in *Drosophila*
William R. Jeffery University of Texas at Austin Austin, Tex.	Analysis of Cytoplasmic Determinants in Ascidian Embryos
John Gerhart University of California, Berkeley Berkeley, Calif.	The Origin of Polarity in the Amphibian Egg
Roger Pedersen University of California, San Francisco San Francisco, Calif.	Cell Fate and Commitment in Mouse Embryo Development
Graham Walker Massachusetts Institute of Technology Cambridge, Mass.	Regulation of Cellular Responses to DNA Damage
Ira Herskowitz University of California, San Francisco San Francisco, Calif.	Determination of Yeast Cell Type
Nina Agabian University of California, Berkeley Berkeley, Calif.	Differential Expression of Trypanosome Surface Antigens

Workshop Summary

This workshop was held to help the committee examine the role of models in the study of development. The participants spoke of research on several different biological systems with which key advances are currently being made. The organisms discussed ranged from bacteria and yeast to worms, frogs, and mice.

Research concerned with development illustrates well the major conclusions embodied in this report. One of these is that biological research concerned with the elucidation of a fundamental mechanism or phenomenon may be carried out -- indeed should be carried out -- with a diversity of organisms. Lower organisms are extraordinarily important in this respect, and the basic knowledge obtained is generally applicable to human biology. Indeed, studies concerned with basic questions in developmental biology have been carried out over the past 100 years with diverse organisms, ranging from bacteria, slime molds, jellyfish, sea urchins, and worms to frogs, chickens, and mice. In these various systems, there are certain common themes and questions. In some cases, the mechanisms responsible may turn out to be the same as in the human; in others, they may be different but nevertheless highly instructive for the general understanding of biological systems and strategies.

Development also illustrates well some of the limitations in transferring information from one group to another, which is more readily done with more general and basic than with specific aspects. Long ago biologists noted that to understand embryology one must study embryos. The bottom line of the argument remains, however: the development of the brain cannot be studied in organisms lacking a nervous system, and knowledge concerning the mechanism of gastrulation is not directly applicable to species in which gastrulation does not occur. Nevertheless, it is evident that basic knowledge in some areas, such as those concerning the control and timing of gene expression in developmental systems, is of central importance to understanding nervous system development -- and knowledge concerning gastrulation contributes to our understanding of morphogenetic movements, which occur in numerous other, and sometimes less readily studied, instances.

William Wood (University of Colorado) described a research strategy involving concerted investigations of a few experimentally favorable model systems, i.e., particular organisms chosen for intensive study. His comments were of special interest to the committee in view of the fact that the worm *Caenorhabditis elegans*, the organism studied by Wood, has been used as a model only within the past decade -- largely because of the insight and efforts of Sydney Brenner (Brenner, 1974; Strome and Wood, 1982; Sulston *et al.*, 1983; Wood *et al.*, 1984). But if the development of new systems is to continue, one cannot dictate which systems are acceptable. Strength in biomedical research relates in part to the continued development of new model systems, an end that cannot be achieved by fiat. Brenner's decision to promote *C. elegans* as a system for the study of development was not based simply on brilliant insight or a good guess; he relied in great measure on decades of study by many biologists, ranging from the systematist to the physiologist and biochemist. Basic research in biology is important to the continued development of the knowledge that serves as a foundation for the selection and development of model systems.

Wood described features of *C. elegans* that make it a favorable model system for studying several aspects of animal development. His studies are directed toward achieving an understanding of the mechanisms by which developmental fates of early embryonic cells are determined in *C. elegans*. The fates appear to be governed by internally segregating, cell-autonomous determinants. As an approach to ascertaining the nature and location of such determinants, Wood and coworkers used laser microsurgery to show that exposing the nucleus of a non-gut-precursor cell to gut-precursor cytoplasm can cause the progeny of the resulting hybrid cell to express gut-specific differentiation markers, indicating that at least one such determinant is cytoplasmic.

To obtain molecular probes for such determinants, Wood's group generated a library of monoclonal antibodies to early embryonic antigens and screened it by immunofluorescence microscopy for antibodies reacting with lineage-specific components (Strome and Wood, 1982). Screening of about 1,500 hybridoma clones has so far yielded five such antibodies, all of which recognize the same component -- cytoplasmic granules (P granules), which segregate specifically with the germ line in early cleavages and are found uniquely in germ line cells throughout the life cycle. P granules -- almost certainly the same structures as those seen by electron microscopy in *C. elegans* and in several other organisms -- are called germinal plasm, nuage, or polar granules.

Experiments on mutant embryos resulting from defects in fertilization or early cleavage and on normal embryos treated with inhibitors of cytoskeletal functions indicate that segregation of the P granules depends on fertilization and requires microfilament action but is largely independent of spindle and microtubule functions. Work on the biochemical nature and function of the P granules is in progress (Wood *et al.*, 1984).

Both the putative gut-lineage determinant and the P granules could be cytoplasmic determinants in the classical sense of maternally synthesized lineage-specific macromolecules that become segregated during early cleavage of the appropriate cells, whose fates they somehow then dictate. However, there is no compelling evidence for this view in *C. elegans*. One alternative possibility for further investigation is that the P granules are a storage form of macromolecules required for germ-cell proliferation. Another is that cell fates are determined through lineage-specific embryonic gene expression, which is controlled by differing concentrations of a few common regulatory molecules in a manner similar to that proposed for other global developmental decisions, such as determination of segment identities in *Drosophila* (Struhl, 1981) or of cell lineage patterns in postembryonic *C. elegans* development (Sternberg and Horvitz, 1984).

Paul Sternberg (University of California, San Francisco) discussed genes controlling cell fate during nematode development. The cell lineage of *C. elegans* is rigidly programmed and invariant. In most cases a cell suffers its usual fate even after destruction of its neighbors (e.g., by a laser microbeam). Those cases in which cell fate is not autonomous involve a small group of multipotential cells that are constrained by their ancestry to a few alternative fates, which are specified by cell-cell interactions. Certain precursor cells are determined to generate identical cell lineages. The execution of such an intrinsically programmed sublineage is indicative of the state of a precursor cell.

Genes that may determine which cell fate prevails have been identified by mutations that cause homeotic transformations in cell fate, i.e., mutations that cause particular cells to adopt the fates normally associated with other cells. Homeotic mutations that have been identified affect both intrinsically and extrinsically determined cell fates, as exemplified by mutants in the genes *lin-17* and *let-23*, respectively. Certain sister cells, normally programmed to have different fates in wild-type, have the same fate in *lin-17* mutants. The same fate is also adopted in *let-23* mutants by a pair of bipotential cells (P11 and P12), which in wild-type have different fates as a result of cell-cell interactions. The homeotic transformations caused by a decrease of *lin-17* or *let-23* gene activity indicate that these genes are necessary for the normal determination of cell fate. However, these data do not imply that such genes necessarily play an instructive role in determination. Rather, their presence may be required only for normal determination. Another locus, *lin-12*, specifies which of two alternative fates several groups of cells will adopt. A high level of *lin-12* activity in a cell causes it to adopt a certain fate, whereas a low level causes it to adopt another. Thus, *lin-12* may control a binary switch that restricts the fates of cells in several sets.

For several other *C. elegans* genes besides *lin-12*, mutations that exert opposite effects on gene activity cause opposite homeotic transformations in aspects of cell fate. Such genes are excellent candidates for controlling cell fate. For example, *tra-1* controls the sexual identity of cells in response to the X:autosome ratio, and *lin-14* controls the temporal identity of many cells.

The *lin-12* locus is also involved in vulva development. Although vulval precursor cells choose among three fates, the level of *lin-12* activity specifies only one binary decision, as evidenced by the effect of both the amount of anchor cell signal and the level of lin-12 activity on vulval precursor cell fates. More than 100 mutations defining more than 25 genes that affect vulval development have been characterized, and the interactions of these mutations with lin-12 mutations have been analyzed. The results are consistent with lin-12 primary involvement in the determination of vulva precursor cell fate.

Gary Struhl (Harvard University) described selector genes that regulate developmental fate in *Drosophila*. He noted that studies of cell lineage and mutations in insects have provided one of the few sources of information about how particular genes control the development of specific parts of an animal's body. This information has led in turn to a concrete model of the genetic logic of animal development called the compartment or selector gene hypothesis (Crick and Lawrence, 1975; Garcia-Bellido *et al*., 1979).

The essence of this hypothesis may be stated as follows. During early development, the insect embryo is subdivided into a series of primordial cell groups, each destined to form a particular part (or compartment) of the larval and adult body. This process of determination involves the selective activation of certain combinations of selector genes in each of these founder groups. Once activated, these combinations of genes remain active in all the descendent cells and specify the developmental pathway that is to be followed. Thus, according to the compartment hypothesis, animals develop in a modular fashion in which a small number of control genes govern the growth and spatial differentiation of many genealogically discrete cell populations.

The formulation of the compartment hypothesis rests in large part on the properties of segments and homeotic genes in *Drosophila*. Insect segments are initially established as a series of adjacent cell groups at the blastoderm stage. Subsequently, all the descendants from each segmental primordium develop as exclusive cell populations, each giving rise to a particular larval and adult segment. Segments are therefore, by definition, compartments (Lawrence, 1981).

Segments also appear to be the precise sites in which a small number of homeotic genes control development. For example, the bithorax complex contains a set of discrete genetic functions that act in certain segments of the thorax and abdomen (Lewis, 1978; Struhl, 1981). None of these genes is active in the head or in the first two thoracic segments; however, beginning in the third thoracic segment and then extending segment by segment throughout the abdomen, first one, then a second, then a third bithorax gene are successively activated up to the eighth abdominal segment, in which all the bithorax genes are active. Thus, all the cells forming each abdominal segment inherit a particular combination of active and inactive bithorax genes. This combination constitutes a code word, or genetic address, that dictates the particular form and cell pattern that normally characterizes each segment.

The key postulate of the compartment hypothesis is that there is a causal link between cell lineage and homeotic gene function -- namely, that the insect body is initially determined by the irreversible activation of particular homeotic genes in the founder cells of each segment. This hypothesis raises many questions: Is it true in detail at the cellular level, e.g., is the realm of activity of a given bithorax gene (or all bithorax genes) precisely that of set segmental compartments? How are the correct number and sequence of segmental primordia established? How are the correct combinations of selector genes activated in each primordium? How do they remain stably active or inactive during the rest of development? How do selector genes dictate a developmental pathway (e.g., how do they control growth and cell pattern within a segment)? And finally, how general is the compartment hypothesis -- is it limited to segmented invertebrates or is it perhaps generally applicable to all animals that have meristic (segmented) body plans? Many of these questions can now be addressed directly, since the two homeotic gene complexes of _Drosophila_ (bithorax and antennapedia) have been cloned (Bender _et al_., 1983; Garber _et al_., 1983). Moreover, a number of other genetic functions that are responsible for segmentation per se or for the heritable activation of the bithorax and antennapedia genes have been identified (and in some cases, also cloned). Indeed, perhaps the most exciting development has been the recent discovery that several homeotic genes in _Drosophila_ include a common amino acid coding sequence called the "homeo-box" (McGinnis _et al_., 1984a; Scott and Wiener, 1984), a small number of which is also found in the genomes of many vertebrates, including frogs, birds, and mammals (McGinnis _et al_., 1984b). These results suggest that many of the principles that govern the development of insect segments might have direct counterparts in vertebrate systems.

William Jeffery (University of Texas at Austin) pointed out that although cytoplasmic determinants in development were described long ago (see Wilson, 1925), the nature, spatial distribution, and mode of action of these substances remain unknown. He elaborated on the value of the ascidian embryo as a model system for the study of such determinants. Ascidian embryos develop rapidly, they have larva consisting of a small number of different cell types, and their mode of development and structure appears to be related to that of vertebrates (Reverberi, 1961). Most significantly, however, some ascidian eggs (i.e., those of _Styela_) contain pigmented cytoplasmic regions that can be traced into individual cell lineages during embryogenenis (Conklin, 1905). These colored markers and other features have been used to generate a description of cell lineages to the larval stage (Nishida and Satoh, 1983; Zalokar and Sardet, 1984).

The cell lineage most suitable for experimental analysis is the one from which larval and adult musculature develops. That lineage appears to be determined at least by the 64-cell stage and is derived almost entirely from eight yellow-pigmented cells located in the vegetal-posterior region of the embryo. The muscle progenitor cells obtain their yellow coloration from a specialized region of the uncleaved egg known as the yellow crescent. That region is also a part of a specialized egg cytoskeletal domain that originates during oogenesis and is maintained in the muscle lineage progenitor cells (Jeffery and Meier, 1983). Initially, it is positioned in the cortex of unfertilized eggs. During ooplasmic segregation it forms a crescent in the vegetal region of the zygotes, and during cleavage it is specifically partitioned to the muscle progenitor cells (Jeffery, 1984). There is some evidence that the cells containing yellow crescent cytoplasm contain factors that can promote muscle cell features in cells lacking the yellow crescent. Whittaker (1980) was able to divert cytoplasm from the yellow-pigmented cells into clear cells by compressing embryos at the third cleavage. Although these embryos subsequently ceased development, some of the clear cells produced acetylcholinesterase, an enzyme normally synthesized only in the muscle cell lineage.

Jeffery's presentation illustrated how ascidian embryos are used to test the theory that cytoplasmic determinants may in part be composed of maternal RNA. He also described how these embryos are used to examine the mechanics of determinant localization and summarized current progress on the use of ascidian embryos to develop physiological assay systems for cytoplasmic determinants.

John Gerhart (University of California, Berkeley) described studies concerned with the origin of polarity in the egg of the African toad *Xenopus* studied by experimental manipulations. For example, the cortical-endoplasmic rotation maybe prevented or artificially restored, the vegetal hemisphere can be activated in several different places, or the blastomeres may be rearranged at the blastula stage (Gerhart, 1980; Gerhart *et al*., 1981; Scharf and Gerhart, 1980, 1983).

Gerhart's results and those of others have elucidated the development of the body axis in the *Xenopus* embryo, especially of the notochord lineage of cells that at one stage induce the formation of the embryo's central nervous system. Their findings indicate that the oocyte, even at the earliest stages of oogenesis, has a polarized organization apparent in the positions of the nucleus, centrosome, Golgi apparatus, and division bridge. This polarity may be directly responsible for the animal-vegetal axis formed by the oocyte in its growth and differentiation. In the full-grown oocyte, there are two primary cytoplasmic localizations important for subsequent development: the animal hemisphere and the vegetal hemisphere. The former contains the nucleus, common organelles, and cortical black pigment; the latter comprises close-packed yolk platelets and no cortical pigment.

At fertilization, the sperm enters at random in the animal hemisphere. The single sperm aster enlarges and moves to the center of the egg. Midway in the first cell cycle, the entire cortex and endoplasm of the egg rotate relative to one another by 30°. The direction of rotation is cued by the sperm aster, although the aster is not needed for the rotation to occur. As a result of the rotation, the grey crescent forms at the juncture of the animal and vegetal hemispheres at the position most distant from the sperm entry point. The notochord lineage of cells normally originates from the grey crescent region of the egg.

As the cortex and endoplasm rotate relative to one another, latent determinative agents are activated in the vegetal hemisphere near the grey crescent. This is a secondary cytoplasmic localization -- a specialization of the primary vegetal localization. This process establishes on the egg a secondary polarity related to the ultimate dorsal-ventral organizatin of the embryo. Any portion of the vegetal hemisphere can become activated to form the localization, although normally only one region does so, depending on the direction of the cortical-endoplasmic rotation.

In the blastula stage, cells cleaved from the vegetal hemisphere induce nearby animal hemisphere cells to undertake a mesodermal type of development. All cells of the animal hemisphere are capable of responding to this induction, although only those close to the vegetal hemisphere normally do so. The strongest induction originates from the region of the secondary localization in the vegetal hemisphere. Above it, animal hemisphere cells are induced to develop toward the notochord. The induction concerns at least the preparations of blastula cells for gastrulation, namely, the types of movement the cells will display and the time at which they begin gastrulation. The most strongly induced cells begin gastrulation first.

The time of gastrulation may determine the initial characteristics of the migrating cells, the duration of migration, the distance travelled, and the types of interactions encountered with other cells. These factors may determine the mesodermal fates of the cells, such as the formation of notochord.

As pointed out by Roger Pedersen (University of California, San Francisco), the mouse embryo has been studied extensively in recent years as a model for mammalian development. As a result, we have a rather complete picture of the timing and morphology of early development and, more importantly, of the prospective potency of early cleavage-stage blastomeres. After fertilization, the embryo cleaves slowly, requiring 20 to 24 hours for the first and second divisions and 10 hours for each of the subsequent cleavage divisions. At the 16-cell stage, many or perhaps all the embryo's blastomeres remain totipotent. The outer cells of embryos in the late cleavage stage differentiate into trophectoderm, whereas the inner cells become the inner cell mass (ICM) (Gardner, 1978).

Studies of experimental chimeras have also provided information about the prospective fate of the trophectoderm and the ICM (Rossant, 1984). The trophectoderm contributes only to the placenta, but the ICM contributes cells to the fetus as well as to the amnion, yolk sac, and the placenta. Thus, according to the mouse model, the location of totipotent cells in late cleavage-stage embryos of placental mammals appears to be a decisive event in the ultimate fate of those cells.

To understand how mouse embryo cells acquire their positions (i.e., how cell allocation to trophectoderm and ICM occurs), Pedersen's group analyzed cell lineage using microinjected horseradish peroxidase and rhodamine-conjugated dextran as tracers (Balakier and Pedersen, 1982). Their results indicate that the ICM is formed by cells that are internalized in the 4th and 5th cleavage divisions. After division to the 32-cell early blastocyst stage, outer cells (now differentiated into trophectoderm) no longer have inner cell descendants. On the other hand, ICM cells contribute to the trophectoderm during blastocyst growth. Trophectoderm cells overlying the ICM are concomitantly displaced from the ICM region. It is not yet clear when the trophectoderm and ICM become clonally distinct populations; however, lineage studies show that the extended potency of cleaving mouse embryo cells is an integral part of normal development, because descendants can cross the boundary between ICM and trophectoderm.

Pederson and collaborators have also studied the fate of horseradish peroxidase-injected cells in gastrula-stage mouse embryos. Their preliminary studies indicate the fate of labeled embryonic endoderm cells and validate the use of this approach for mapping the fate of cells at the threshold of organogenesis. The anterior movement of labeled cells implicates the primitive streak as a source of proliferating cells during gastrulation and formation of the embryonic axis of bilateral symmetry.

These studies demonstrate that cell lineage can be analyzed in the mouse embryo _in vitro_ and provide important information about cell fate in the intact embryo. The ICM and the primitive streak appear to serve as sources of cells during embryo growth. The results have interesting implications for the uses of mice or other vertebrate embryos for the study of mammalian development.

As discussed above, models for the study of some aspects of development need not be confined to embryological systems. In particular, the temporal control of gene expression is of key importance, and valuable models for the study of the mechanism involved may be found in other biological systems. Three such systems, and very different ones (bacteria, yeast, and trypanosomes), were described in the presentations at the workshops.

Graham Walker (Massachusetts Institute of Technology) reviewed studies in bacteria to examine cellular responses to DNA damage. These studies have provided insight into some of the mechanisms that are involved in the modulation of the expression of global regulatory networks of genes (Walker, 1984).

Exposure of _E. coli_ to agents or conditions that either damage DNA or interfere with DNA replication elicits a set of diverse physiological responses termed the SOS responses. These include the induction of prophage, the induction of an increased capacity to repair UV-irradiated phage, the induction of a capacity for the cells to be mutated by ultraviolet light and various other mutagens, filamentous growth as a result of septum inhibition, a temporary cessation of respiration, and other responses. The SOS responses are coordinately regulated by the products of two genes, Rec and LexA. The mechanisms used to regulate this set of responses are of particular interest with respect to developmental models, since the cell transiently turns on a set of functions that are coordinately regulated but that in many cases bear no obvious physiological relationship to each other.

For many years, the study of SOS network regulation was complicated by both the complexities of the responses and the interrelationships of the key regulatory elements. However, the application of powerful genetic and molecular biological techniques has provided a detailed picture of the regulation of this complex network. The network is now known to consist of more than 17 genes, each of which is repressed by the LexA protein. Induction of the genes in the SOS network occurs when the RecA protein becomes activated in response to a signal generated by DNA damage. Once activated, RecA mediates the cleavage of LexA. A number of the mechanisms by which the cell fine tunes the responses of individual genes in the network are also understood. At least some of the SOS-inducing agents also induce a separate regulatory network governing the expression of the heat-shock responses.

Exposure of *E. coli* to methylating and ethylating agents induces another set of responses termed the adaptive responses, which may have parallels in some mammalian cells. Studies of the regulation of this network have given us insights into a unique control circuit, whose key positively acting element, the Ada protein, contains methyltransferase, which repairs three different DNA lesions. Thus the key regulatory protein not only controls its own expression and that of other genes in the network but also functions as a major induced repair protein.

Analysis of mutants affecting a specific process can lead to the identification of the normal components of that process, especially genes that control that process in some manner. Several genes identified by genetic analysis may play significant roles in determining cell type. These include the classic bithorax complex of *Drosophila* (Lewis, 1978) and the major sex-determining locus *tra-1* of nematodes (Hodgkin, 1980).

Yet another example of such a regulatory phenomenon is found in the mechanism whereby yeast cell type is determined. This system was summarized by Ira Herskowitz (University of California, San Francisco). Three cell types occur in *Saccharomyces cerevisiae*: a, α, and a/α. The a and α cells are specialized for mating; the a/α cell, formed by mating between a and α cells, is specialized for meiosis and sporulation. A single genetic locus, the mating type locus (MAT), is responsible for these behaviors: cells with the MATα allele exhibit the α mating type; cells with the MATa allele exhibit the a mating type; and cells with both MATa and MATα exhibit the a/α cell type. These gross phenotypic differences are based on biochemical specializations that play physiological roles in the mating process. In particular, a and α cells produce characteristic mating pheromones, a-factor and α-factor, respectively, which specifically cause arrest of cells of the opposite cell type in the G-1 phase of the cell division cycle. The a/α cells, which do not mate with any cell type, neither produce nor respond to the pheromones. The biological purpose of the mating factors appears to be coordination of the cell division cycles of mating partners to facilitate mating. In addition to the pheromone systems, other cell specializations include agglutination between a and α cells, production of an α-factor inactivation activity by a cells, and a difference in the budding pattern between a or α cells and a/α cells (Herskowitz, 1983).

Because the mating type locus determines all the properties described above, it has been proposed that this locus contains regulatory products that govern expression of many other genes, distinct from MAT (Mackay and Manney, 1974b, Strathern *et al.*, 1981). Genes that might be regulated in this fashion were identified by MacKay and Manney (1974a), who isolated mutants with defective mating -- so-called *ste* (sterile) mutants.

From this and later work, it is evident that in yeast there is a single locus responsible for determining cell type. Numerous genes scattered throughout the genome are concerned with the specializations of the three yeast cell types. The expression of these genes is regulated by the proteins encoded by the mating type locus. Herskowitz pointed out that a key question in cellular differentiation is whether there are analogous master regulatory loci in higher eucaryotes. Genetic analysis of *Drosophila* and of nematodes will almost certainly lead to the observation of such genes. As quoted by Herskowitz:

> The genome contains not only a series of blue-prints, but a co-ordinated program of protein synthesis and the means of controlling its execution (Jacob and Monod, 1961).

Nina Agabian (University of California, Berkeley) described current research concerning an even more complex regulatory phenonomenon -- the variable expression of surface antigen in a parasitic African trypanosome. This is a remarkable case, involving the sequential and exclusive expression of a single variable surface glycoprotein (VSG) gene from a repertoire of hundreds of genes. The appearance of new immunodominant glycoproteins in parasite populations is facilitated by immune selection and provides a mechanism for evading the host immune system and manifesting the disease condition (Parsons *et al.*, 1984a).

Agabian and her collaborators developed a working hypothesis that accounts for two basic features of the system: (1) VSG gene expression is regulated at the level of transcription and (2) VSG genes and flanking sequences are capable of genomic rearrangement (Buck *et al.*, 1984). Specific position effects seem to be required for gene activation. In every case studied, the transcriptionally active VSG gene is located near a telomere and is embedded in a region of repetitive DNA. Placement of VSG genes in telomeric sites is necessary but not sufficient for transcriptional activation, since many VSG genes retain their telomeric position but are inactivated either during antigenic variation or during parasite development.

Gene duplication and transposition and extended regions of gene conversion or deletion have all been implicated in the variety of genomic rearrangements correlated with antigenic variation. Perhaps one purpose of these rearrangements is the placement of a VSG gene in a telomeric environment.

In general, VSG genes occur in gene families, which contain 1 through 10 or more members. Many of the genes are located in chromosome internal sites; however, estimates suggest that at least 25% of the VSG genes in the cell occupy telomeric regions. Thus a picture emerges in which the trypanosome genome contains many VSG genes whose genomic environment is compatible with or prerequisite to gene activation.

Since a number of potentially active telomeric VSG genes exist, what ensures that only a single VSG gene is transcribed? One potential clue was provided by the observation that all VSG mRNAs had a common 5' terminal untranslated leader sequence (Boothroyd and Cross, 1982). Agabian proceeded on the assumption that this 35 nucleotide spliced leader (SL) might reveal the 5' boundary of the expression site (Nelson *et al.*, 1983). Using a synthetic oligonucleotide complementary to the SL, it was determined first that the SL is not encoded by the DNA immediately flanking VSG structural genes and second that these sequences are reiterated approximately 200 times in the genome of *Trypanosoma brucei*, mostly in a large tandem array of directly repeated 1.4 kb units, each containing a single SL exon.

Of several hypotheses that attempt to explain the presence of the SL at the 5' terminus of trypanosome mRNAs, Agabian favors the one that the SL is added to these transcripts by an intermolecular process. The two simplest mechanisms for such a process are primed transcription and posttranscriptional splicing, or ligation. Both models predict that a discrete SL containing RNA is transcribed from the SL-reiteration units of trypanosomatid species (Parsons *et al.*, 1984b).

The role of SL in gene expression must be defined in order to understand the mechanisms of transcription and of the processing of RNA in these organisms. One must first determine whether the SL is important in priming mRNA transcription and, thus, participates in the formation of the nascent transcript. The existence of the SL, however, can no longer be used to explain the transcriptional exclusivity of VSG gene expression. The challenge remains to understand what special constellation of factors results in the selection of a single telomeric VSG gene for transcriptional activation.

In sum, this workshop provided the committee with a comprehensive overview of the ways that lower organisms can serve as sources of systems for the modelling and critical examination of key questions concerning biological mechanisms in development. The invited participants demonstrated the importance of in-depth studies of a given system but, at the same time, illustrated the value of having a repertoire of different systems as well as the value of developing and introducing new ones.

REFERENCES

Balakier, H., and R. A. Pedersen. 1982. Allocation of cells to inner cell mass and trophectoderm lineages in preimplantation mouse embryos. Dev. Biol. 90:352-362.

Bender, W., M. Akam, F. Karch, P. A. Beachy, M. Peifer, P. Spierer, E. B. Lewis, and D. S. Hogness. 1983. Molecular genetics of the bithorax complex in *Drosophila melanogaster*. Science 221:23-29.

Boothroyd, J. C., and G. A. Cross. 1982. Transcripts coding for variant surface glycoproteins of *Trypanosoma brucei* have a short, identical exon at their 5' end. Gene 20:281-289.

Brenner, S. 1974. The genetics of *Caenorhabditis elegans*. Genetics 77:71-94.

Buck G. A., S. Longacre, A. Raibaud, U. Hibner, C. Giroud, T. Baltz, D. Baltz, and H. Eisen. 1984. Stability of expression-linked surface antigen gene in *Trypanosoma equiperdum*. Nature 307:563-566.

Conklin, E. G. 1905. The organization and cell-lineage of the ascidian egg. J. Acad. Nat. Sci. Phila. 13:1-119.

Crick, F. H., and P. A. Lawrence. 1975. Compartments in insect development. Science 189:340-347.

Garber, R. L., A. Kuroiwa, and W. J. Gehring. 1983. Genomic and cDNA clones of the homeotic locus in antennapedia in *Drosophila*. EMBO J. 2:2027-2036.

Garcia-Bellido, A., P. A. Lawrence, and G. Morata. 1979. Compartments in animal development. Sci. Amer. 241:102-110.

Gardner, R. L. 1978. The relationship between cell lineage and differentiation in the early mouse embryo. Pp. 205-241 in W. J. Gehring, ed. Genetic Mosaics and Cell Differentiation. Springer-Verlag, Berlin.

Gerhart, J. C. 1980. Mechanisms regulating pattern formation in the amphibian egg and early embryo. Pp. 133-316 in R. Goldberger, ed. Biological Regulation and Development, Vol. 2. Plenum Press, New York.

Gerhart, J., G. Ubbels, S. Black, K. Hara, and M. Kirschner. 1981. A reinvestigation of the role of the grey crescent in axis formation in *Xenopus laevis*. Nature 292:511-516.

Herskowitz, I. 1983. Determination of yeast cell type. Symp. Soc. Dev. Biol. 41:65-75.

Hodgkin, J. 1980. More sex-determination mutants of *Caenorhabditis elegans*. Genetics 96:649-664.

Jacob, F., and J. Monod. 1961. Genetic regulatory mechanisms in the synthesis of proteins. J. Mol. Biol. 3:318-356.

Jeffery, W. R. 1984. Pattern formation by ooplasmic segregation in ascidian eggs. Biol. Bull. 166:277-298.

Jeffery, W. R., and S. Meier. 1983. A yellow crescent cytoskeletal domain in ascidian eggs and its role in early development. Dev. Biol. 96:125-143.

Lawrence, P. A. 1981. The cellular basis of segmentation in insects. Cell 26:3-10.

Lewis, E. B. 1978. A gene complex controlling segmetation in *Drosophila*. Nature 276:565-570.

MacKay, V., and T. R. Manney. 1974a. Mutations affecting sexual conjugation and related processes in *Saccharomyces cerevisiae*. I. Isolation and phenotypic characterization of nonmating mutants. Genetics 76:255-271.

MacKay, V., and T. R. Manney. 1974b. Mutations affecting sexual conjugation and related processes in *Saccharomyces cerevisiae*. II. Genetic analysis of nonmating mutants. Genetics 76:273-288.

McGinnis, W., M. S. Levine, E. Hafen, A. Kuroiwa, and W. J. Gehring. 1984a. A conserved DNA sequence in homeotic genes of the *Drosophila* antennapedia and bithorax complexes. Nature 308:428-433.

McGinnis, W., R. L. Garber, J. Wirz, A. Kuroiwa, and W. J. Gehring. 1984b. A homologous protein-coding sequence in *Drosophila* homeotic genes and its conservation in other metazoans. Cell 37:403-408.

Nelson, R. G., M. Parsons, P. J. Barr, K. Stuart, M. Selkirk, and N. Agabian. 1983. Sequences homologous to the variant antigen mRNA spliced leader are located in tandem repeats and variable orphons in Trypanosoma brucei. Cell 34:901-909.

Nishida, H., and N. Satoh. 1983. Cell lineage analysis in ascidian embryos by intracellular injection of a tracer enzyme. I. Up to the eight-cell stage. Dev. Biol. 99:382-394.

Parsons, M., R. G. Nelson, and N. Agabian. 1984a. Antigenic variation in African trypanosomes: DNA rearrangements program immune evasion. Immunol. Today 5:43-50.

Parsons, M., R. G. Nelson, K. P. Watkins, and N. Agabian. 1984b. Trypanosome mRNAs share a common 5' spliced leader sequence. Cell 38:309-316.

Reverberi, G. 1961. The embryology of ascidians. Adv. Morphogen. 1:55-101.

Rossant, J. 1984. Somatic cell lineages in mammalian chimeras. Pp. 89-109 in A. McLaren and N. LeDouring, eds. Chimeras in Developmental Biology. Academic Press, London.

Scharf, S. R., and J. C. Gerhart. 1980. Determination of the dorsal-ventral axis in eggs of Xenopus laevis: Complete rescue of UV-impaired eggs by oblique orientation before first cleavage. Dev. Biol. 79:181-198.

Scharf, S. R., and J. C. Gerhart. 1983. Axis determination in eggs of Xenopus laevis: A critical period before first cleavage, identified by the common effects of cold, pressure, and ultraviolet irradiation. Dev. Biol. 99:75-87.

Scott, M. P., and A. J. Weiner. 1984. Structural relationships among genes that control development: Sequence homology between the Antennapedia, Ultrabithorax, and fushi tarazu loci of Drosophila. Proc. Natl. Acad. Sci. USA 81:4115-4119.

Sternberg, P. W., and H. R. Horvitz. 1984. The genetic control of cell lineage during nematode development. Ann. Rev. Genet. 18:489-524.

Strathern, J., J. Hicks, and I. Herskowitz. 1981. Control of cell type in yeast by the mating type locus: The alpha 1- beta 2 hypothesis. J. Mol. Biol. 147:357-372.

Strome, S., and W. B. Wood. 1982. Immunofluorescence visualization of germ-line-specific cytoplasmic granules in embryos, larvae, and adults of Caenorhabditis elegans. Proc. Natl. Acad. Sci. USA 79: 1558-1562.

Struhl, G. 1981. A gene product required for correct initiation of segmental determination in *Drosophila*. Nature 293:36-41.

Sulston, J. E., E. Schierenberg, J. G. White, and J. N. Thomson. 1983. The embryonic cell lineage of the nematode *Caenorhabditis elegans*. Dev. Biol. 100:64-119.

Walker, G. C. 1984. Mutagenesis and inducible responses to deoxyribonucleic acid damage in *Escherichia coli*. Microbiol. Rev. 48:60-93.

Whittaker, J. R. 1980. Acetylcholinesterase development in extra cells caused by changing the distribution of myoplasm in ascidian embryos. J. Embryol. Exp. Morphol. 55:343-354.

Wilson, E. B. 1925. The Cell in Development and Heredity. 3rd ed. Macmillan, New York. 1232 pp.

Wood, W. B., E. Schierenberg, and S. Strome. 1984. Localization and determination in early embryos of *Caenorhabditis elegans*. Pp. 37-49 in E. H. Davidson and R. Firtel. eds. Molecular Biology of Development. UCLA Symposia on Molecular Cellular and Biology, New Series, Vol. 19. Alan R. Liss, New York.

Zalokar, M., and C. Sardet. 1984. Tracing of cell lineage in embryonic development of *Phallusia mammillata* (Ascidia) by vital staining of mitochondria. *Dev. Biol.* 102:195-205.

Appendix G

Mathematical Modelling in Biomedical Research

Workshop Program
October 4, 1984
Washington, D.C.
Kenneth B. Bischoff
Chairman and Organizer

Arthur C. Guyton University of Mississippi Medical Center Jackson, Miss.	Mathematical Modelling of Control of the Circulatory System
George Bell Los Alamos National Laboratory Los Alamos, N. Mex.	Mathematical Models in Cellular Immunology
Richard Bergman University of Southern California Medical Center Los Angeles, Calif.	The Minimal Model Approach and Its Application to Aging and Metabolic Disease
Garrett O'Dell Rensselaer Polytechnic Institute Rensselaer, N.Y.	Mathematical Modelling in Development: How the Slime Mold Slug Crawls
Robert Dedrick National Institutes of Health Bethesda, Md.	Pharmacokinetics of Regional Drug Delivery
Charles DeLisi National Institutes of Health Bethesda, Md.	Mathematical Modelling of Molecular Structure-Function Relationships

Mathematical models were considered in several of the preceding workshops, but in some it was not feasible or suitable. To provide a broader view of how mathematical modelling can aid in the understanding of biomedical systems, a separate session was held. Of course, much of biophysical chemistry could be considered as a mathematical description of fundamental processes, but a somewhat more macroscopic level was considered appropriate for most of this workshop.

Much of the work to date has involved processes in higher animals, especially mammals, but the specific topics addressed in this session ranged from protein structure and cellular immunology to development and to physiologic and pharmacologic processes in mammals. One common thread in all the presentations was the interplay between good mathematical modelling and experiment. Modelling in a vacuum generally does not lead to results as useful as those involving attempts to describe or predict specific experimental results. The theoretical model may also suggest further experimentation.

According to Arthur Guyton (University of Mississippi Medical Center), mathematical modelling has been applied to physiological systems for decades. One of the first was the circulatory system, beginning in the 19th century (Attinger, 1964). Early attempts were based on simple graphic methods of analysis, which were subsequently replaced by analog and digital computer techniques that have resulted in models with hundreds of variables. This evolution can be exemplified by the work of Guyton and colleagues in their studies on the regulation of blood pressure (Guyton, 1955; Guyton and Coleman, 1969; Guyton *et al*. 1972.

The use of computers to explore very large models has led to the discovery of exciting new concepts. Several of these discoveries have been made in the field of hypertension:

- When first developed, the computer model of arterial pressure regulation demonstrated that the feedback loop for the renal regulation of blood volume and arterial pressure mechanism has very high sensitivity (infinite gain), but it is highly modulated (damped). Because of this infinite gain, the kidney-fluid volume mechanism for controlling arterial pressure can override essentially all other pressure controlling mechanisms for long-term control of arterial pressure.

- Contrary to long-standing belief, quantitative analysis of renal function with a large computer model has predicted that even vast increases in tubular reabsorption of sodium will, by itself, have relatively little effect on arterial pressure, because the increased pressure results in acceleration of glomerular filtration rate, which can rapidly override the increased sodium reabsorption. This prediction has been borne out in recent experiments demonstrating that massive stimulation of sodium reabsorption by the administration of very large amounts of aldosterone causes only about a 15 mm Hg rise in arterial pressure.

- Ischemic kidneys secrete large amounts of renin, which causes the formation of angiotensin in the circulating blood. Angiotensin in turn constricts the peripheral blood vessels, and the animal develops hypertension. Therefore, most investigators have believed that this peripheral increase in resistance is the <u>cause</u> of the hypertension. Computer models have demonstrated, however, that increasing the resistance in all the body's blood vessels except in the kidneys will not cause hypertension. This implied that the hypertension resulting from the angiotensin must be related in some way to the effect of angiotensin on the kidney. In subsequent experiments, angiotensin infused directly into the renal arteries was observed to cause severe hypertension because of its effects on the kidneys, which caused them to retain sodium and water.

- Most hypertensive persons do not have a significant increase in extracellular fluid volume, blood volume, or cardiac output. Thus, many researchers believe that volume has little to do with the control of arterial pressure. Complex computer models have predicted that in the early stages of hypertension there will be increased fluid volumes as well as increased cardiac output and, thus, excess blood flow through the peripheral tissues. Over time these increases will cause an increase in total peripheral resistance, which will then exert a negative effect on the body fluid volumes and on cardiac output, returning both to levels so nearly normal that the usual measuring techniques cannot detect the abnormality. Animal experiments have demonstrated these sequential changes. Thus, basic physiological experiments have been quantitatively refined and explained by mathematical models. Relatively complex computer models were required to define the more subtle interactions, which were then verified by the experiments suggested by the models.

Richard Bergman (University of Southern California Medical Center) discussed glucose intolerance associated with the progression of various debilitating conditions that are prevalent in the aging population, including heart disease, kidney disease, and blindness. It is of great interest to understand the pathogenesis of glucose intolerance in the aged, and its association with other age-related conditions such as obesity (Chen <u>et</u> <u>al</u>., 1985).

The ability to dispose of a carbohydrate load depends on several specific factors: the ability of glucose to stimulate insulin secretion, the sensitivity of the tissues to insulin, and the ability of glucose to accelerate its own uptake by cells, independent of insulin. To understand the pathogenesis of glucose intolerance, these factors must be measured; yet, it has required extensive experimental invasion to evaluate them *in vivo* (Bergman and Cobelli, 1980; Bergman *et al*., 1981b; Toffolo *et al*., 1980).

Modelling has been an important tool for measuring important metabolic parameters. One approach is to use the digital computer to analyze the dynamic plasma insulin and glucose patterns observed after glucose ingestion. The computer finds the parameters of two simple, or minimal, models: one of insulin kinetics (including ß-cell secretion and insulin metabolism) (Bergman and Cobelli, 1980; Finegood *et al*., 1984) and one of glucose kinetics (including the effects of insulin and glucose on glucose utilization) (Bergman *et al*., 1981a). In the course of fitting the model to data, five characteristic parameters of metabolic function can be determined from dynamic data. These parameters, which constitute an integrated metabolic profile, can be used to compare normal and pathological states, including aging.

This approach, used with minimal models, has been used to examine metabolic changes in lean, aged individuals and to compare them with similar indices in young, obese subjects. Although increased adiposity is a result of the aging process, the metabolic changes of the aging individual can be differentiated from the obese. Both groups are resistant to insulin; however, in young, obese, nondiabetic persons, pancreatic responsiveness increases to compensate for the resistance; in obese elderly persons, such compensation fails to occur. In older people, therefore, the primary defect may be an inability of the ß-cells to effectively increase insulin output in response to the challenge of insulin resistance.

Metabolic defects in aging people have been attributed simply to their well-documented diminished carbohydrate intake. Studies with altered diets have revealed, however, that although increased carbohydrate intake affects the ß-cell responsivity of young and old subjects, the resultant effect on the peripheral tissues differs. In the older subjects, but not in the young, the insulin sensitivity of peripheral tissues was improved by a high carbohydrate intake, which increased pancreatic function. In fact at 80% carbohydrate in the diet, the insulin sensitivity (and glucose tolerance) of the older subjects is not different from the young. Therefore, high carbohydrate diets equalize the insulin sensitivity of old and young subjects. These results are consistent with a ß-cell defect being the primary cause of glucose intolerance in the aged (Chen *et al*., 1985).

The separation of the various possibilities was feasible only through the use of mathematical computer analysis, leading to the quantitative metabolic profile parameters. In this case, the minimal, or not very complicated, model was deemed adequate for the level of sophistication sought.

Robert Dedrick (National Institutes of Health) discussed the pharmacokinetics of targeted drug delivery. Anticancer drugs are generally not very selective. That is, their toxicity to sensitive normal tissues may approach or even exceed their toxicity to the tumor. The resulting limited therapeutic advantage has prompted many investigators to consider pharmacologic approaches to improving drug effectiveness. Through regional drug administration, it may be possible to expose tumor-bearing tissue to drug concentrations substantially exceeding exposures of normal tissues (Collins, 1983; Dedrick et al., 1978, 1984; Fenstermacher et al., 1981; Flessner et al., 1984). Technical advances, such as special-purpose pumps, including implantable catheters and access systems, have helped to further the clinical application of this type of biological administration (Molnar et al., 1980). The development of sophisticated imaging techniques (which are also based on mathematical methods) often enable the clinician to monitor tumor response to therapy (Groothuis et al., 1984).

Pharmacokinetic models for intracavitary and intraarterial drug administration are well established. It is frequently possible to use mathematical models to predict the pharmacokinetic advantage of a particular drug, schedule, and route of administration with precision sufficient for clinical guidance.

Some major unresolved issues relate to the spatial distribution of the pharmacokinetic processes (Collins and Dedrick, 1983; Collins et al., 1982). The uniformity of concentration within the cavity or infused tissue is not known, but one might suspect drug maldistribution in many cases. Although lumped models may predict, for example, that the fluid adjacent to the tumor in the peritoneal cavity has a much higher drug concentration than systemic plasma, the drug must be transported through the tumor to reach subsurface cells. This leads to distributed reaction-diffusion models, which are of great interest but which are confounded to a large degree by unknown tumor morphology and transport and metabolic characteristics. Similarly, infused tumors may exhibit considerable and unpredictable heterogeneity in both morphology and relevant transport parameters (Groothuis et al., 1984).

These important theoretical and experimental issues require extension of pharmacokinetic theory beyond conventional lumped models and pose both challenges and opportunities concerning its applicability to individual clinical situations. Precise definition of the complex reaction-diffusion processes will require increasingly sophisticated mathematical approaches.

This is just one recent advance in applied pharmacokinetics. A comprehensive text by Gibaldi and Perrier (1982) provides many other examples.

In developmental biology, a basis for mathematical modelling can be found in the following concept. If genes endow embryonic cells with certain general, plausible, mechanochemical response properties, then those properties, iterated many times by interacting cells, can lead inexorably, without further genetic intervention, to complex globally coordinated tissue movement behavior. Because of this amplification scheme, genes need specify far less information to build an organism and direct its behavior than would be required in a conventional "blueprint." Mathematics is a language available for deducing which large-scale collective consequences flow from which microscale response properties conferred by genes on the low-level participants.

This general idea is illustrated by a case study with the cellular slime mold Dictyostelium discoideum, which was described by Garrett O'Dell (Rensselaer Polytechnic Institute). Mathematical reasoning, based on standard ideas from classical continuum mechanics, suggested that there was only one mechanochemically feasible method by which known behavior of individual D. discoideum slime mold amoebae constituting the pseudoplasmodium could result in propulsion of the organisms. However, the proposed propulsion method was inconsistent with much of the experimental literature. That conflict has motivated collaborative experimentation in somewhat new directions. Although the predictions of the new modelling have not yet been completely verified through experimentation, these attempts to model the long-studied and still mysterious phenomenon of slug locomotion have already served to stimulate work that may lead eventually to real understanding, which is all any mathematical model is supposed to do.

The new model for the proplusion of the D. discoideum pseudoplasmodium grex is based on mechanics: how does each participating amoeba contribute motive force, and how do the many force contributions produce a coordinated collective effort? When a pseudoplasmodium migration is stalled by mechanically arresting the motion of its boundary, the amoebae in it actively circulate in a reverse fountain flow extending the length of the body. The velocity of individual cells relative to the grex boundary is commensurate with the migration speed of a grex: approximately one grex length per hour. Further experiments have indicated that this cell fountain flow is not an artifact resulting from the stalling of its progress. Clark and Steck (1979) and Durstin and Vork (1979) argue that cell circulation constitutes the propulsive engine of migrating pseudoplasmodia.

More precisely, Sternfeld and David (1981) have postulated that each participating amoeba orients its attempted motion by the same cyclic adenosine monophosphate (cAMP) chemotaxis used during aggregation. The cAMP concentration field within the grex consists of pulses that are emitted periodically at the tip and propagate toward the rear by the same cAMP relay behavior seen during aggregation (Schaap and Wang, 1984). Chemotactic migration within pseudoplasmodia described by Durstin and Vork (1979) generally reinforces this description. The principal new contribution of this model is its resolution of the following apparent conceptual difficulty: in a close-packed three-dimensional mass of cells, each amoeba trying to crawl can exert traction only on its neighbors, which in turn exert traction on it. Without a rigid agar substratum to crawl upon, equal efforts by a cohort of amoebae to crawl in the same direction, each upon similarly crawling neighbors, cancel and produce no net mechanical result. Gradients of other chemicals must arise naturally within the grex, approximately perpendicular to the average cAMP gradient. If one such chemical modulates the traction amoebae exert individually, then a self-regulating, chemotactically oriented fountain flow of cells must result. A further result of this reasoning is that slight modification of the chemotactic behavior needed for aggregation can account also for the migratory behavior of the grex and erection of the fruiting body.

As discussed by George Bell (Los Alamos National Laboratory), the fact that mathematical models are not universally recognized in the field of immunology is due partly to the very complexity of the system, whose elements and essential interactions are subject to much disagreement. Under such conditions, mathematical models are best able to play a heuristic role, stimulating lines of inquiry and sharpening conclusions that can be drawn rather than providing large-scale definitive theories.

A major theme in modern immunology concerns the importance of idiotypes and their recognition by other elements of the immune system, as embodied most elegantly in Jerne's idiotypic network, whose formulation was motivated in part by early mathematical models of the immune system (Bell, 1970, 1971) and their failure to predict a self-limiting immune response (Jerne, 1974, 1976, 1984). Network models have in turn stimulated a large body of experimental immunology (e.g., Bona and Kohler, 1983). Sometimes the models are purely conceptual; sometimes they are quite mathematical (e.g., Gunther and Hoffmann, 1982; Hiernaux, 1977). In both cases, however, they have guided investigators in the design and interpretation of experimental programs.

There is little agreement about the overall importance of idiotypic interactions in immune responses. Nevertheless, well-chosen mathematical models can be of great help in clarifying the logical options. Even very simple models may help. For example, Jerne (1984) has argued that the small numbers of observed antibody-secreting cells, together with antibody concentrations in blood, imply a very long lifetime of many antibodies and, hence, a relatively stable network with a long memory. More complex models may be even more helpful. At present, many computer scientists are studying the self-organizing behavior of complex systems. Some are currently formulating model immune networks that have approximately 10^4 interacting elements. Their goal is to understand by computer simulation the qualitative dynamics of such systems -- their stability and evolution in response to external changes (antigens) or to internal perturbations such as new clones. The results of such studies may have important implications for immunology.

Nearly every cellular immunologist has a favored scheme for regulating a response by T-cells of various types -- helpers, suppressors, or contrasuppressors. Many of the more deserving of these schemes might be greatly clarified by a modest amount of mathematical modelling and computer simulation.

As a quite different example, models of cell-cell interaction are useful in cellular immunology. For example, investigators have analyzed conjugates containing various numbers of cytotoxic T lymphocytes (CTL) and target cells. From the kinetics of delivering a lethal signal to the target and the subsequent expression of that signal (Perelson and Macken, in press), one can conclude that a CTL affects only one target cell at a time. This sort of analysis is now being used to guide an interdisciplinary team studying mechanisms of the generation and action of cytotoxic lymphocytes.

Various other examples of modelling can be found in studies of cell-cell or cell-substrate interactions. For example, Bell *et al*. (1984) have shown that when adhesion is mediated by the formation of intercellular bridges between specific receptors on the cells, there is a strong tendency for such bridges to become highly concentrated in limited areas of intercellular contact, forming specialized membrane areas and structures. Receptor redistribution has been observed in experimental studies on phagocytosis (Michl *et al*., 1983) and is being examined for many other systems.

In the Soviet Union, mathematical models of the immune response are presumably used as a basis for treatment in clinical medicine. Of particular importance in clinical applications are the mathematical models of the antiviral immune response developed by Marchuk and Petrov (1983), who took into account both cellular and humoral effects and included macrophages, T helper cells, T effector cells, ß cells, and antibodies. The mathematical properties of the models have been analyzed, and the models have been used to treat pneumonia patients (Marchuk *et al*., 1983).

Charles DeLisi (National Institutes of Health) discussed mathematics and computers in molecular biology. He observed that the number of sequenced nucleotides exceeds 3 million and is currently tripling approximately every 2 years. Recognition that computers play an important role in the routine processing and manipulation of these data -- e.g., restriction enzyme mapping, gel assembly, signal sequence identification -- is widespread (Soll and Roberts, 1984). The extraordinarily rapid advances in computer technology hold promise for progress in other areas, including the prediction of molecular structure and the functional and positional classification of deduced protein sequences. Thus, the technology may be applied to a wide range of problems spanning fundamental questions about the structure, assembly, and regulation of genes; the function of receptors and other membrane proteins; and the formulation of biologically active peptides and drugs.

During the past decade, a number of mathematical scientists have developed pattern recognition algorithms that rapidly search large data bases for sequence similarities (Lipman and Wilbur, 1983; Smith and Waterman, 1981). These algorithms are now used routinely when new sequences are determined. They are especially useful when protein function is not known, which is often the case when a sequence is deduced from the genetic code. Insights have been especially fruitful in the area of oncogenes. It was recently discovered that the transforming agent of the simian sarcoma retrovirus is strongly homologous to human platelet-derived growth factor -- more than 80% of the residues matching. Soon afterward, investigators discovered an equally strong homology between the erythroblastoma ß retrovirus (erbB) oncogene product and the epidermal growth factor receptor.

Although these discoveries strengthen and lend precision to existing concepts, one feature of the erbB homology is puzzling: the epidermal growth factor receptor phosphorylates tyrosines, whereas the homologous region on the erbB product does not appear to share similar activity. The difference might be the result of an inadequate biochemical assay, or it might reflect the well-known fact that a change in even a *single* residue, let alone 15% of the residues, can have a pronounced effect on function. These observations remind us that the gap between sequence and function is large, and that spanning it is crucial to understanding, and ultimately to controlling, cell regulation.

Two types of advances will be especially important to deepening our understanding of the structure-function relationship: (1) the ability to scan a data base for three-dimensional rather than one-dimensional patterns -- a difficult problem that is expected to become more pressing as cloning stimulates the accumulation of crystallographic data -- and (2) the ability to accurately predict higher order from lower order structure -- a need that arises from the enormous number of molecules whose structures must be learned for experimental and theoretical progress and from the slowness of the experimental methods for determining them.

In addition to drawing functional conclusions from sequence properties, we would like to deduce the cellular location of the protein product. Precise prediction will again probably depend on detailed knowledge of higher order structure, but we may be able to go a large part of the way in many instances on the basis of sequence alone. In particular, we consider generic positional categories: nucleic acid binding proteins, cytoplasmic proteins, membrane proteins, and so on, and within each category, more specific locations.

The simplest distinction is between integral and peripheral membrane proteins. A number of investigators have developed methods based on some aspect of hydrophobicity, the most precise reported by Klein *et al.* (in press). Their method can be used with very few errors -- two out of 160 proteins allocated. Futhermore, it can be used to calculate the odds of correct classification and to determine the boundaries of membrane-spanning segments to within experimental error. These allocations to integral or peripheral categories are based on information that the sequence is membrane-associated. The more important and difficult problem is to make an unconditioned allocation.

The analysis of sequences for patterns that are more complex than just blocks of hydrophobic and hydrophilic residues is important to understanding both the location and possible activity of peptide sequences (Guy, 1984; Kaiser and Kezdy, 1984). Thus sequences are generally passed through a Fourier analyzer to detect periodicities in properties, such as in hydrophobicity, or in charge. By failing to perform such an analysis, some important observations can be missed. For example, the acetylcholine receptor passes back and forth through the membrane several times. A hydrophobicity analysis indicates that each of the five receptor subunits has four membrane-spanning segments. If the receptor, as currently believed, is capable of channel formation, the model with membrane embedded hydrophobic segments makes this difficult to visualize. However, when the sequence is subjected to Fourier analysis, it can be seen that an amphipathic segment is located between the third and fourth hydrophobic blocks. Guy (1984) has postulated that the five homologous amphipathic regions, one from each subunit, aggregate to form a cylindrical membrane-spanning channel, whose interior is composed of the polar faces of each of the segments and whose exterior, in contact with lipid, is composed of the apolar faces. The model makes a number of unambiguous predictions, including a carboxyl terminus on the exterior of the membrane.

Attempts at more detailed predictions of the structure of small and intermediate size peptides are also in progress, most notably in the studies by Pincus et al. (1983). Free energy minimization calculations on mellitin have predicted locations of alpha carbon atoms that are in close accord with those determined by x-ray crystallography. Applications to the P21 protein product of the EJ bladder carcinoma cell have been successful in showing how a single base substitution can lead to a marked conformational change in the gene product.

A great deal more effort is needed to generalize this technique to other types of structures in polar environments. The limiting factor in such an effort is not the availability of computers. Rather, it is the development of a better understanding of the theoretical principles that control molecular folding in a complex environment. That understanding can be achieved only through a commitment to the training and support of theoretical scientists.

REFERENCES

Attinger, E. O., ed. 1964. Pulsatile Blood Flow. McGraw-Hill, New York. 462 pp.

Bell, G. I. 1970. Mathematical model of clonal selection and antibody production. J. Theor. Biol. 29:191-232.

Bell, G. I. 1971. Mathematical model of clonal selection and antibody production, II. J. Theor. Biol. 33:339-378.

Bell, G. I., M. Dembo, and P. Bongrand. 1984. Cell adhesion: Competition between nonspecific repulsion and specific bonding. Biophys. J. 45:1051-1064.

Bergman, R. N., and C. Cobelli. 1980. Minimal modeling, partition analysis and the estimation of insulin sensitivity. Fed. Proc. 39:110-115.

Bergman, R. N., C. R. Bowden, and C. Cobelli. 1981a. The minimal model approach to quantification of factors controlling glucose disposal in man. Pp. 269-296 in C. Cobelli and R. N. Bergman, eds. Carbohydrate Metabolism: Quantitative Physiology and Mathematical Modeling. John Wiley & Sons, Inc., New York.

Bergman, R. N., L. S. Phillips, and C. Cobelli. 1981b. Physiological evaluation of factors controlling glucose tolerance in man: Measurement of insulin sensitivity and B-cell glucose sensitivity from the response to intravenous glucose. J. Clin. Invest. 68:1456-1467.

Bona, C. A., and H. Kohler, eds. 1983. Immune Networks. Ann. N. Y. Acad. Sci. 418:1-395.

Chen, M., R. N. Bergman, G. Pacini, and D. Porte, Jr. 1985. Pathogenesis of age-related glucose intolerance in man: Insulin resistance and decreased ß-cell function. J. Clin. Endo. Metab. 60:13-20.

Clark, R. L., and T. L. Steck. 1979. Morphogenesis in Dictyostelium: An orbital hypothesis. Science 204:1163-1167.

Collins, J. M. 1983. Pharmacokinetics of intraventricular administration. J. Neurooncol. 1:283-291.

Collins, J. M., and R. L. Dedrick. 1983. Distributed model for drug delivery to CSF and brain tissue. Am. J. Physiol. 245:R303-R310.

Collins, J. M., R. L. Dedrick, M. F. Flessner, and A. M. Guarino. 1982. Concentration-dependent disappearance of fluorouracil from peritoneal fluid in the rat: Experimental observations and distributed modeling. J. Pharm. Sci. 71:735-738.

Dedrick, R. L., C. E. Myers, P. M. Bungay, and V. T. DeVita, Jr. 1978. Pharmacokinetic rationale for peritoneal drug administration in the treatment of ovarian cancer. Cancer Treat. Rep. 62:1-11.

Dedrick, R. L., E. H. Oldfield, and J. M. Collins. 1984. Arterial drug infusion with extracorporeal removal. I. Theoretic basis with particular reference to the brain. Cancer Treat. Rep. 68:373-380.

Durstin, A. J., and F. Vork. 1979. A cinematographical study of the development of vitally stained Dictyostelium discoideum. J. Cell Sci. 36:261-279.

Fenstermacher, J. D., R. G. Blasberg, and C. S. Patlak. 1981. Methods for quantifying the transport of drugs across brain barrier systems. Pharmac. Ther. 14:217-248.

Finegood, D. T., G. Pacini, and R. N. Bergman. 1984. The insulin sensitivity index: Correlation in dogs between values determined from the intravenous glucose tolerance test and the euglycemic glucose clamp. Diabetes 33:362-368.

Flessner, M. F., R. L. Dedrick, and J. S. Schultz. 1984. A distribution model of peritoneal-plasma transport: Theoretical considerations. Am. J. Physiol. 246:R597-R607.

Gibaldi, M., and D. Perrier. 1982. Pharmacokinetics, 2nd ed. Marcel Dekker, New York. 494 pp.

Groothuis, D. R., P. Molnar, and R. G. Blasberg. 1984. Regional blood flow and blood-to-tissue transport in five brain tumor models: Implications for chemotherapy. Prog. Exp. Tumor Res. 27:132-153.

Gunther, N., and G. W. Hoffmann. 1982. Qualitative dynamics of a network model of regulation of the immune system: A rationale for the IgM to IgG switch. J. Theor. Biol. 94:815-855.

Guy, H. R. 1984. A structural model of the acetylcholine receptor channel based on partition energies and helix packing calculations. Biophys. J. 45:249-261.

Guyton, A. C. 1955. Determination of cardiac output by equating venuous return curves with cardiac response curves. Physiol. Rev. 35:123-129.

Guyton, A. C., and T. G. Coleman. 1969. Quantitative analysis of the pathophysiology of hypertension. Circ. Res. 24 (Suppl.): 1-19.

Guyton, A. C., T. G. Coleman, and H. J. Granger. 1972. Circulation: Overall regulation. Ann. Rev. Physiol. 34:13-46.

Hiernaux, J. 1977. Some remarks on the stability of the idiotypic network. Immunochem. 14:733-739.

Jerne, N. K. 1974. Clonal selection in a lymphocyte network. Pp. 39-48 in G. M. Edelman, ed. Cellular Selection and Regulation in the Immune Response. Raven Press, New York.

Jerne, N. K. 1976. The immune system: A web of V-domains. Harvey Lectures, Ser. 70:93-110.

Jerne, N. K. 1984. Idiotypic networks and other preconceived ideas. Immunol. Rev. 79:5-24.

Kaiser, E. T., and F. J. Kezdy. 1984. Amphiphilic secondary structure: Design of peptide hormones. Science 223:249-255.

Klein, P., M. Kanehisa, and C. DeLisi. In press. Detection and classification of membrane-spanning proteins. Biochem. Biophys. Acta

Lipman, D. J., and W. J. Wilbur. 1983. Contextual constraints on synonymous codon choice. J. Mol. Biol. 163:363-376.

Marchuk, G. I., and R. V. Petrov. 1983. The mathematical model of the anti-viral immune response. Pp. 161-173 in G. I. Marchuk and L. N. Belykh, eds. Mathematical Modeling in Immunology and Medicine. North Holland, New York.

Marchuk, G. I., V. M. Zhivoderov, A. P. Berbentsova, L. G. Tambortseva, K. G. Astahova, and K. S. Agafonnikova. 1983. Biological stimulation of immunotherapy of acute pneumonia patients. Pp. 343-349 in G. I. Marchuk and L. N. Belykh, eds. Mathematical Modeling in Immunology and Medicine. North Holland, New York. pp.

Michl, J., M. M. Pieczonka, J. C. Unkeless, G. I. Bell, and S. C. Silverstein. 1983. Fc receptor modulation in mononuclear phagocytes maintained on immobilized immune complexes occurs by diffusion on the receptor molecule. J. Exp. Med. 157:2121-2139.

Molnar, P., D. Broothuis, R. Blasberg, D. Zaharko, E. Owens, and J. Fenstermacher. 1984. Regional thymidine transport and incorporation in experimental brain and subcutaneous tumors. J. Neurochem. 43:421-432.

Perelson, A. S., and C. A. Macken. In Press. Quantitative models for the kinetics of cell-mediated cytotoxicity at the single cell level. In P. Henkert and E. Martz, eds. Mechanisms of Cell-Mediated Cytotoxicity II. Plenum Press, New York.

Pincus, M. R., J. van Renswoude, J. B. Harford, E. H. Chang, R. P. Carty, and R. D. Klausner. 1983. Prediction of three-dimensional structure of the transforming region of the EJ/T24 human bladder oncogene product and its normal cellular homologue. Proc. Natl. Acad. Sci. USA 80:5253-5257.

Schaap, P., and M. Wang. 1984. The possible involvement of oscillatory cAMP signaling in multicellular morphogenesis of the cellular slime molds. Dev. Biol. 105:470-478.

Smith, T. F, and M. S. Waterman. 1981. Identification of common molecular substances J. Mol. Biol. 147:195-197.

Soll, D., and R. J. Roberts, eds. 1984. The Application of Computers to Research on Nucleic Acids. Part I, pp. 1-428; Part II, pp. 429-856. I.R.L. Press, Oxford, United Kingdom.

Sternfeld, J., and C. N. David. 1981. Cell sorting during pattern formation in _Dictyostelium_. Differentiation 20:10-21.

Toffolo, G., R. N. Bergman, D. T. Finegood, C. R. Bowden, and C. Cobelli. 1980. Quantitative estimation of ß-cell sensitivity to glucose in the intact organism: A minimal model of insulin kinetics in the dog. Diabetes 29:979-990.

Index

B

C